MODERN PHYSICS
Notes, Problems & Solutions

Philip A. Stahl

(M.Phil., Physics)

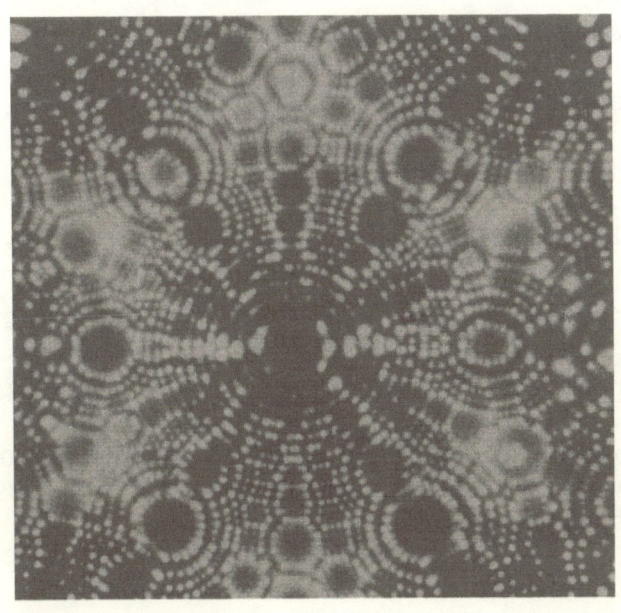

Table of Contents

ISBN: 978-1-304-08361-6

Preface

The material in this text grew out of class lessons, problems and notes for special (and general) relativity and quantum mechanics while teaching Calculus Physics, Relativity and Quantum Mechanics at different institutions from 1986 through 1992.

This is a revision of my earlier 2014 text, whereby I have replaced the 'Parts' (and multiple sub-chapters_ with single, separate chapters- for greater clarity. There are also many more end of chapter problems and selected solutions at the end of the text. A more extensive Mathematical Supplement has also been included featuring new material on vector analysis and differential equations.

As with my two earlier texts: ***Physics Notes for Advanced Level***, and ***Astronomy & Astrophysics: Notes, Problems and Solutions***, this text is intended more as a survey course dealing with specific topics deemed common to modern physics. It is not intended to be a comprehensive text, say on the order (or scale!) of Robert Martin Eisberg's *Fundamentals of Modern Physics*.

Still, there is a heavy emphasis on problem solving, and more than 55 pages of problem solutions. Hopefully students will attempt the many end of chapter problems – which are intense and variegated - independently before consulting the solutions.. If diligent students can work the problems (all or most of them herein) on their own they ought to be well prepared for more advanced courses in modern physics, or quantum mechanics.

3

Chapter I. Special Relativity

1.1. Michelson-Morley Experiment.

In the popular mind, at least, the word "relativity" usually conjures up visions of space travelers returning youthful to Earth after journeys of many decades (measured on Earth). Of course, there is much more to relativity than this. As a matter of fact many aspects of relativity as in the case of motions, aren't new at all. Basically, it merely entails the assertion that the laws of physics appear to be the same in given reference frames or coordinate systems. This is made more accurate in Einstein's special relativity by referring to *inertial reference frames.*

Though Henri Poincare came close to discovering the principles of relativity, it was Albert Einstein first and foremost who ruthlessly and relentlessly pursued the basic principles to their utterly logical conclusions, including that lengths shrink as velocities tend toward the speed of light, c, and time slows or "dilates".

Like Ernst Mach before him, Einstein adopted the view that in considering two objects in relative motion, it is futile and meaningless to attempt to decide which object is "really" in motion and which is at rest. If you are in a Jumbo jet traveling at a speed of 900 km/hr relative to the Earth, it makes no difference whether someone says you are moving at that speed, or the Earth is moving at that speed. In either case, the operation of the laws of physics in your jet and on the Earth will be the same. Balls will still drop, and you will still fall off your chair if not careful. There exists no absolute frame of reference.

The Special Theory of Relativity, interestingly, may be said to have had its origins in the null result of an experiment the primary aim of which was to detect relative motion. This experiment was first carried out by the American physicist Albert A. Michelson in 1881, and subsequently repeated in 1887, with the help of Edward W. Morley. The experiment has thus come to be known as "the Michelson-Morley experiment".

Michelson-Morley experiment   1887

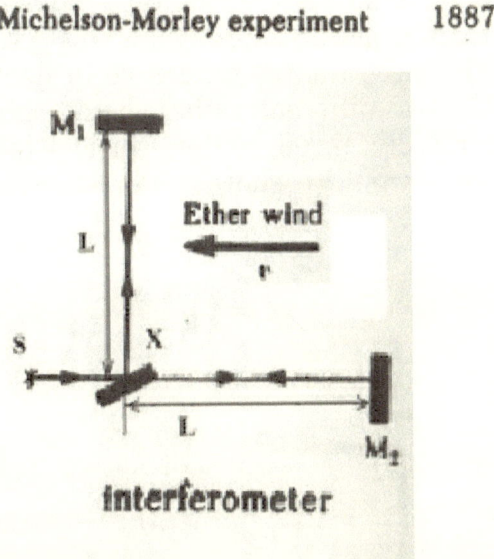

interferometer

Fig. 1.1. Michelson-Morley –Experiment

The basic idea was to time the transits of light in two distinct directions: perpendicular to the Earth's orbital motion, and parallel to the orbital motion. This was to be accomplished by using an arrangement of mirrors and light beams such as depicted in Fig. 10.1. A difference in light velocities (transit distance/ time) would reveal itself by a delicate interference pattern formed by two separate beams after rejoining each other.

The implication of a difference in velocities would mean the confirmation of a remarkable entity called "the Ether". In effect, if light represents waves propagating through the Ether, the velocity of light as recorded by instruments on Earth's surface must be distorted by the motion of the Earth through space. An analogous principle is that a swift river must retard a swimmer's combined upstream and downstream speeds more than his cross-current speed. Similarly, a large difference in light velocities (along the two different paths) should show Earth is moving rapidly through the Ether, while a small difference would show it's moving slowly

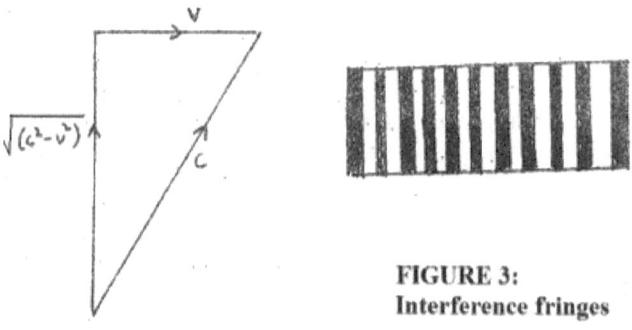

FIGURE 3:
Interference fringes

FIGURE 2

Astonishingly, on each occasion the experiment was conducted the result was virtually negative. There was no evidence of any Ether flowing past the Earth in any direction. Any minor deviations fell within the purview of the experimental errors.

Imagine the apparatus in Fig. 1 to be moving with velocity v toward the right, then only in the event of *a null result* should there should there also be a relative velocity of the Ether of magnitude v to the left. To get a positive result the apparatus needs to move a speed v relative to the Ether. The round trip time for a light beam following path X-M2-X is:

$$t_1 = 2Lc/ (c^2 - v^2)$$

Meanwhile, the light beam traveling from X to M1 must have a component of velocity v along X-M2, relative to the hypothesized Ether or it will not strike the mirror at M1. Since the velocity of the light relative to Ether is c, subtracting the preceding component leaves a velocity of $(c^2 - v^2)$ e.g. see Fig. 2. The same is true for the return journey to the total time t2 for the path X-M1-X is:

$$t_2 = 2L/ (c^2 - v^2)^{1/2}$$

The recombination of the two light beams produces interference fringes as depicted in Fig. 3. Any difference in the times to traverse the paths shows up as a shift in the position of the bright and dark fringes indicating different path lengths.

Hence, a difference in time Δt was found equivalent to a path difference: $c (\Delta t)/ \lambda (d)$ where d = the width of one fringe and lambda is the wavelength of the light (taken to be 6×10^{-7} m). From this, the time difference between the two beams could be computed from:

$$\Delta t = t_1 - t_2 = 2L/c\{ 1/ (1 - v^2/c^2) - 1/ (1 - v^2/c^2)^{1/2}\}$$

This equation can be approximately expressed as:

$$\Delta t \approx L/c \, (v^2/c^2)$$

Since $v \ll c$, this corresponds to a fringe shift of:
$$\Delta d - c \, (\Delta t)/ \lambda = L \, c^2/ \lambda \, (c^2).$$

The measurement of the fringe shift is accomplished by rotating the whole apparatus through 90 degrees. The effect of this is to interchange the arms, X-M1 and X-M2, thereby reversing the sign of the fringe shift so that an overall shift of 2 (delta d) should be observed. For the Michelson -Morley experiment of 1887, L = 11 m, v = 0.0001c. Using the path difference equation for Δ d, we arrive at: Δ d = 0.183 or, approximately 0.2 fringe. The overall shift expected was: 2 x (Δ d) = 2 x (0.2) = 0.4 fringe But the largest shift *actually detected* was only 0.01 fringe within the experimental error.

This null result flabbergasted physicists of the time. They were simply unable to conceive that light required no medium within which to propagate. Hence, it's not difficult to see why so many clung to the Ether McGuffin for so long, even after it was disproven. Thus, fanciful and elaborate schemes were thought up to explain the null result, much like today's intelligent design proponents have confected fanciful explanations to try and disavow Darwinian evolution.

For his own part, Michelson naturally assumed that the local ether had to be adhering to the Earth, travelling with it through space. All other scientists were incredulous that an orbital velocity of 30 km/s in relation to the Sun could not generate the tiniest ether "breeze". To many, the situation was not unlike a ship maintaining constant speed and direction in the sea, irrespective of current changes.

1.2 The Lorentz -Fitzgerald Transformation:

In a last ditch effort to satisfactorily explain the riddle of the null result (from the Michelson-Morley experiment), a fantastic idea was put forward by George F. Fitzgerald in 1890. Using the analogy of a rubber ball which is deformed upon striking a wall, Fitzgerald conceived that the ether would distort matter.

This distortion would take the form of a contraction of length in the direction of the motion through the ether. Such a contraction would explain the null result of the Michelson -Morley experiment. That is, the arm L, of the apparatus, moving against the ether would be shortened by "ether pressure" just enough to compensate for the slowing down of light by the ether wind.

A similar but more mathematical theory was worked out by the physicist Hendrik A. Lorentz, who expressed the length contraction in the direction of motion by:

$$L' = L[1 - v^2/c^2]^{1/2}$$

This famous equation became known as the "*Lorentz-Fitzgerald contraction*".

Understandably, the hypothesis of Lorentz and

Fitzgerald gradually gained general acceptance, except for a Swiss Patent clerk, who refused to be deluded by such a contrived idea (based as it was on absolute motion). The clerk's name was Albert Einstein, born in Ulm, Bavaria, in 1879.

Einstein had graduated as a physics major, but he managed to produce such a poor impression as a teacher that he was dismissed from three teaching jobs in succession. Having been reduced to a 'hand-to-mouth' existence, he was lucky to find a job processing patent applications in Bern, Switzerland. Fortunately, he also has a leisurely schedule that permitted him to while away a good many hours contemplating space, time and energy. (A good thing his time was nearly a century ago, otherwise - today-he'd be accused of "gold bricking"!)

After much thought, Einstein was forced to conclude that motion is never observable as motion with respect to space and that there is no basis for the introduction of "absolute motion". In Einstein's mind, the only kind of motion was relative rest perceived from different viewpoints (e.g. "reference frames"). Einstein called this insight the Principle of Relativity. According to Einstein's Principle of Relativity:
"All the laws of physics are the same in all inertial reference frames."

It's instructive at this point to commence a quantitative approach to see exactly how Einstein reasoned. This necessitates we first get accustomed to the idea of a coordinate system. Basically, the purpose of any coordinate system is to enable us to identify a particular point in space. The standard procedure (for

3-dimensional space) is to take three mutually perpendicular axes with coordinates in x, y and z, representing distanced to where the axes meet at the origin, or o. In special relativity such coordinate systems are typically designated: S, S', S" etc. or alternatively, S1, S2, S3.

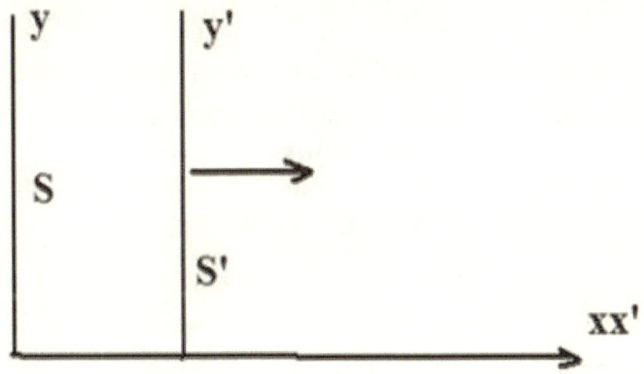

Fig. 1.4. Coordinate systems S and S' in relative motion

In Fig. 4 we have two coordinate systems, S and S', with S having coordinate x, y and S' having coordinates x', y'. (We confine the systems to 2 dimensions here for initial simplicity.) Thus, S' has its origin at o' and S has its own at o. In Fig. 1, S' is shown moving with constant velocity v in the x - x' direction (which obviously coincide for S and S').

Though we arbitrarily assigned S' as moving this depends on which coordinate system is taken to be at rest. An observer attached to system S may very well consider himself moving and S' stationary. This would resemble the well-known example of a passenger in a stationary train observing a train parallel to his through his window and deducing he is moving, though it is actually the train parallel to his. The key point is that the systems S and S' have a constant relative velocity. Such coordinate systems occupy a special place in relativity and are called "inertial reference frames" or "inertial coordinate systems". Their primary feature is that they lack and acceleration of one to the other.

Now, think about this carefully: if the origins O and O' coincide at time t = 0, and we are observers in S', then we will see O' move along OX with velocity v. Then the two sets of coordinates, representing the same point (in S and S') are related by:

$x' = x - vt$ and $y' = y$

If we chose we could append a 3rd axis, i.e. z coming out of the page, and have also:

$z' = z$

The preceding equations would describe the "Galilean transformation". From these it's easy to obtain velocities by differentiating with respect to t:

$dx'/dt = dx/dt = v$

$dy'/dt = dy/dt$

$dz'/dt = dz/dt$

(Note here that we never once wrote t' = t. Why not?)

We start with the preceding transformations and

see how Einstein reasoned. Reference may be made to Fig. 2 which displays a three-dimensional perspective of the relative motion and hence is a bit more complicated. Our aim in using it is to find an alternative to the Galilean transformations - which obviously can't be correct for all situations.

For example, according to the Galilean transformations, a pulse o flight sent out from S' would move with the velocity c, the speed of light, as measured from S', but with the velocity (c + v) measured from S. But this directly contradicts the demonstrated fact (Michelson-Morley) that the speed of light is always constant when computed from one reference frame relative to another.

Einstein thus began afresh by using not only the Principle of Relativity (just stated, i.e. the laws of physics are the same in all *inertial* reference frames). But   also:

The speed of light is always found to have the same value no matter what the motion for the source or the observer.

From these two postulates, Einstein deduced a number of surprising results which would have been totally unacceptable to a more conservative mind.

Start then with the two systems depicted in Fig. 5 which are coincident in space at the instant a flash bulb (say) is set off when the origins coincide. The observer S sees S' moving in the x-direction with the velocity v and observer S' sees S moving with the same velocity in the opposite (-x) direction.

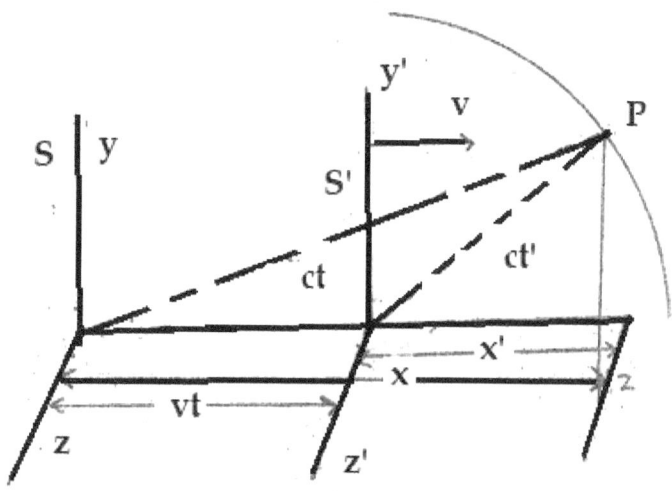

Fig. 1.5. Relative motions of S and S' referred to space-time event P

Both observers see the flash of light travel away from the origin with the same velocity c. The distance the light travels in the S system during time t, is ct. This must be true in any direction. If the light spreads out equally in all directions, then by the end of time t it has extended to fill a sphere whose radius is r. We can write:

(1) $r^2 = x^2 + y^2 + z^2 = c^2 t^2$

The exact same equation holds in the S' system since the flash was set off when O coincided with O. Thus, while the 2 systems are separating from each other, the observer in S' also sees the light fill a spherical shell in his own system, so writes (for r'):

(2) $r'^2 = x'^2 + y'^2 + z'^2 = c^2 t'^2$

Note the velocity of light is the only thing that's the same in both systems. The two preceding equations thus describe a point lying on the spherical shell. Even though they describe the same point in space, the observer in S sees the point at position (x, y, z, t) and the observer S' at position (x', y', z', t').

Subtracting equation (2) from equation (1) and transposing terms:

$$x^2 + y^2 + z^2 - c^2 t^2 = x'^2 + y'^2 + z'^2 - c^2 t'^2$$

We now look for a transformation similar to the Galilean transformation, but which will allow c to be the same in both S and S'. Since the y and z coordinates of the position are not affected by the motion in the x-direction we can say y' = y and z' = z. For the x-coordinate, we try a transformation of the form: x = a(x' + vt') and x = a(x - vt), where a is an invariant determined by the two fundamental postulates (i.e. the same quantity is used in going from x to x' as from x' to x).

Further, we expect a to depend on the velocity v in such a way that it becomes equal to 1 when v becomes very small compared with the speed of light. When this happens, the x and x' transformations become the same as the ordinary Galilean transformations.

So we begin by using x = a(x' + vt') to solve for t' and obtain:

(3) t' = 1/v (x/a - x')

For x' above, we now insert the value for x' (e.g. x' = a(x- vt)):

(4) $t' = 1/v(x/a - ax - avt) = at - x^2(a^2 -1)/ va$

Similarly, we find for t:

(5) $t = -at' + x'(a^2 - 1)/ va$

If we now substitute x' = a(x - vt) and equation (4) into the right hand side of equation (2), we obtain:

(6) $x^2 + y^2 + z^2 - c^2 t^2 = a^2(x - vt)^2 + y^2 + z^2$

$= c^2[at - x/v (a^2 - 1)^2/a]$

Re-arrange terms and cancel the z and y terms, which are the same on both sides of the equation, to get:

(7) $x^2 - c^2 t^2 = [a^2 - (a^2 - 1)c^2/ a^2v^2]x^2$

$+ 2[(a^2 - 1)c^2/v^2 - a^2] xvt - (c^2 - v^2)a^2t^2$

 If the preceding is to hold true for any value of x and t, each term on the left side must equal each term on the right. Since there are no terms with the combination xt on the left, the xt term on the right must be zero. This means:

(8) $(a^2 - 1)c^2/ v^2 - a^2 = 0$

Solving for a:

(9) $a^2 = 1/ (1 - v^2/c^2)$ and $a = [1/ (1 - v^2/c^2)]^{1/2}$

Finally:

(10) $(a^2 - 1)/a = v^2/c^2/ [1/ (1 - v^2/c^2)]^{1/2}$

Substituting the preceding into our x, x' transformation equations and equation (3), we arrive at the following transformations to replace the Galilean:

(11)

$$x = x' + vt'/(1 - v^2/c^2)]^{1/2}$$

and

$$x' = x - vt/ (1 - v^2/c^2)]^{1/2}$$

(12) $y = y'$ and $y' = y$

(13) $z = z'$ and $z' = z$

(14) $t = t' + x'v/c^2/[(1 - v^2/c^2)]^{1/2}$

and

$$t' = t - xv/c^2/[(1 - v^2/c^2)]^{1/2}$$

Equations (11)- (14) are known as the **Lorentz transformation**. Note the important feature is that the *time* must be given as well as the position, because the respective clocks in S and S' will cease to read identical times after they have parted from one another. This is the significance of (14) in the above set. Also, the fact that time is given the same importance as space (i.e. as another dimension) shows there's nothing special or mystical about "the fourth dimension".

Let's now consider the implications of the Lorentz-Einstein transformation and refer again to the original diagram (Fig. 5). Recall:

(11) $x = x' + vt'/(1 - v^2/c^2)^{1/2}$

and

$x' = x - vt/(1 - v^2/c^2)^{1/2}$

(12) $y = y'$ and $y' = y$

(13) $z = z'$ and $z' = z$

(14) $t = t' + x'v/c^2/(1 - v^2/c^2)^{1/2}$

and

$t' = t - xv/c^2/(1 - v^2/c^2)^{1/2}$

Suppose there's a clock located in system S' and an observer in system S sees this clock moving with velocity v. At any time t, the position of the clock with reference to the S −observer's system is given by $x =$ vt. If the length of time between two ticks is of this clock is T' in the S' system, the transformation to the S system (assuming x' = 0) makes the time interval T appear to be: $T = T'/(1 - v^2/c^2)^{1/2}$

Since T > T' then it seems to Observer S his clock is running slower than it does to S'. This applies not only to the clock but to all physical processes that depend on time, e.g. the vibrations of electrons in atoms, rates of chemical reactions, biological processes (heart beat, respiration) etc. In effect it appears to the observer in S that his counterpart in S' is living at a slower rate than he is. However, to the observer in S' it is the observer in S who seems to be living at a slower rate.

This is a paradoxical result but one which can't be escaped if we carry special relativity to its logical conclusion as Einstein did. So long as the two systems have a constant relative motion, we cannot say that one is moving and the other standing still, or that one's clock is moving slowly and the other quickly.

A number of subtle consequences arise out of this. For one thing, the notion of *simultaneous events*, i.e. for observers in two different reference frames, is no longer tenable. Einstein, in fact, seems to have been the first human being to recognize that simultaneity between two events is a provincial illusion. In his landmark (1905) paper on special relativity (*"Does the Inertia of a Body Depend on Its Energy Content?"*) he remarked:

"We have to take into account that all our judgments in which time plays a part are always judgments of simultaneous events. If, for instance, I say 'That train arrives here at 7 o'clock' I mean something like this: 'The pointing of the small hand of my watch to seven o'clock and the arrival of the train are simultaneous events."

Einstein agreed that observing such simultaneity was a reasonably accurate way for a person holding a watch to tell the time of an event happening **next to the watch**, but insisted that on principle the method couldn't be relied upon for timing an event far away from the watch, especially by someone moving in relation to the other things involved.

As an illustration of a type of experiment that attempts to reckon simultaneity, consider Fig. 6 in which a light pulse is directed at a moving mirror M traveling with velocity v to the right. By the time the pulse reaches the mirror it has moved a distance D = v (Δ t)/2 horizontally.

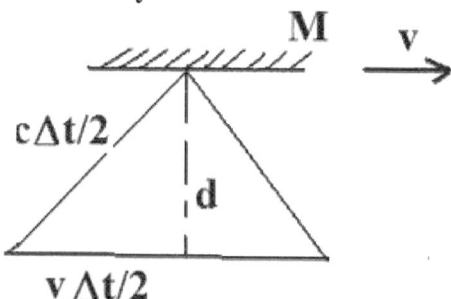

Fig. 1.6. Basic simultaneity experiment.

According to an outside observer, if light is to strike the mirror it must depart at an angle to the vertical as shown. By Pythagoras' theorem (square of the hypotenuse equals the sum of the squares of the other 2 sides): $(c (\Delta t)/2)^2 = (v (\Delta t)/2)^2 + d^2$
This implies: $\Delta t = 2d/(c^2 - v^2)^{1/2} = 2d/ c[(1 - v^2/c^2)^{1/2}]$

But, for the stationary observer:

$\Delta t' = 2d/ c$

which implies:

$\Delta t = \Delta t'/ (1 - v^2/c^2)^{\frac{1}{2}}$

We call Δt the "proper time" i.e. that time interval between two events as measured by an observer who sees the events at the same place. It's always the time measured by a single clock at rest in the frame.

Worked Problem: Consider 3 galaxies: **A**, **B** and **C**.

An observer in **A** measures the velocities of **B** and **C** and finds they are moving in opposite directions - each with a speed of 0.7c relative to him., i.e.

(0.7c)<----------(B)-----(A)-----(C)--------->(0.7c)

What is the speed of **A** observed by someone in **B**?

What is the speed of **C** observed by someone in **B**?

The observer in A thinks that the two other galaxies are receding from him at a rate 1.4c. Show him how this is wrong, by providing the correct result.

Solution:
The speed of **A** observed in **B** = 0.7c, exactly equal to the speed of **B** observed in **A**, by principle of relative velocities.

To find the speed of **C** observed in **B**, we use relativistic addition of velocities or:

$$u = (u' + v)/ 1 + u'v/c^2$$

$$u = (0.7c + 0.7c)/ [1 + (0.7c)(0.7c)/c^2]$$

$$u = (1.4c)/ 1 + 0.49 = 1.4c/1.49 = 0.94c$$

1.3: The Lorentz – Fitzgerald Contraction.

Having dealt with time in the context of special relativity, we now consider what happens to length. We return to the diagram of systems S and S' moving relative to each other (Fig. 7) and consider a meter stick of length L (= 1 m) pointed in the +x direction and moving in that direction with velocity v.

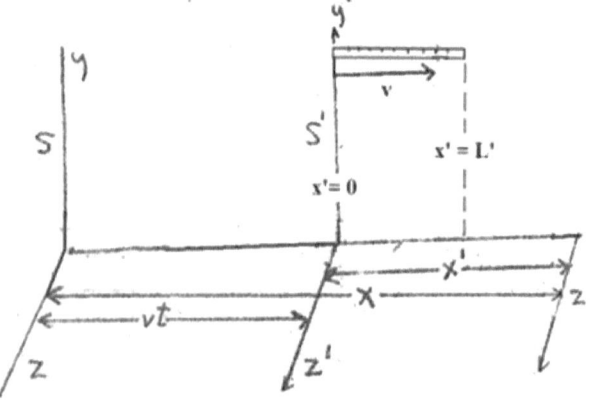

Fig. 1.7. Illustration of Lorentz-Fitzgerald contraction

We can also think of it as being at rest in the S' system with one end at x' = 0 and another end at x' = L', initially.

Its length in the system S' is therefore L'. At time t = 0 we take a photograph of the meter stick with a camera located in the S system. We inquire: What are the positions of the two ends of the meter stick in the S system? We can answer this by using the first of the Lorentz-Einstein transformations:

$$x = x' + vt'/(1 - v^2/c^2)^{1/2}$$

and

$$x' = x - vt/ (1 - - v^2/c^2)^{1/2}$$

We opt to use the lower form rather than the upper, having already specified an instant at t = 0 and hence looking for simplification. Then:

$$x' = x / (1 - v^2/c^2)^{1/2}$$

Now, the end of the meter stick that is at $x' = 0$ in S' is found to be at $x = 0$ in S. The *other* end in S is seen to be at:

$$x = L' (1 - v^2/c^2)^{1/2}$$

This informs us the length of the moving meter stick in S so we can say that the length of the moving meter stick as seen by the camera is:

$$L = L' (1 - v^2/c^2)^{1/2}$$

As can readily be seen, this implies *length contraction*. For example, say the rod is moving at v = 0.6c, then the length L is (given L' = 1m):

$$L = (1 \text{ m}) (1 - (0.6c)^2/c^2)^{1/2} = 1m [(1 - 0.36c^2/c^2)^{1/2}$$

$$L = (1m) (0.64)^{1/2} = 1m (0.8) = 0.8 \text{ m (or 80 cm)}$$

(Remember that by symmetry arguments of relativity, the observer in the other system will argue that $L' = L (1 - v^2/c^2)^{1/2}$ so that *from his viewpoint* a meter stick will be similarly foreshortened.)

The reader may recall that something similar occurred in the time transformation, i.e. each of two relatively moving clocks ran slower with respect to the other than to itself.

This brings us back to the solution of the Michelson-Morley experiment discussed in an earlier chapter. It is the curious, symmetric relativity in time and length (in direction of motion) which solves the paradox of the "missing ether wind" quite simply and logically. Moreover it is solved more fundamentally and satisfactorily than either Fitzgerald or Lorentz could manage.

Let's derive the new law for addition of velocities. Assume an object in the S' system starts at the point x' = 0 at time t' = 0. It moves with constant velocity u' (relative to S') and in the time t' it travels a distance x'.

By definition, $u' = x'/t'$

We ask: 'How fast does this object travel according to the observer at rest in System S?'

This observer will see a velocity given by the formula:

$u = x/t = (x' + vt')'/t' + x'v/c^2$ or

$u = (t' + v)/(1 + x'v/t' c^2)$

$u = (u' + v)/1 + u'v/c^2$

This is *the relativistic formula for addition of velocities.* If u' and v = c then the formula yields:

$u = (c + c)/1 + c^2/c^2$

$u = 2c/1 + 1 = 2c/2 = c$

If u' and v *are less than c* then u must always be less than c.

Worked Problem:

Say two objects are moving at 3c/4 towards each other, then what is their relative velocity as recorded in a system S observing their approach?

Solution:

We have: u' = 3c/4 and v = 3c/4

then:

$u = (u' + v)/1 + u'v/c^2 = (3c/4 + 3c/4)/1 + u'v/c^2$

where: $(3c/4 + 3c/4) = 3c/2$

and:

$u'v = (3c/4)(3c/4) = 9c^2/16$

Then we proceed as follows:

$$u = (3c/2)/ [1 + 9c^2/16/c^2]$$

$$u = (3c/2)/ (1 + 9/16) = (3c/2)/ (25/16)$$

$$u = (3c/2) (16/25) = 24c/25$$

1.4 The Inertia of Energy

The inertia of energy (or light), is embodied in Einstein's most famous equation:

$$E = m c^2$$

which is more accurately posed as:

$$E = (\Delta m) c^2$$

where Δm is the "mass defect" or difference, say in a nuclear reaction, and c is the velocity of light.

Before looking at examples, it's useful to consider the relativistic mass of a particle, in terms of its rest mass m_o. The rest mass, as the term implies is the mass of the object at rest or:

$$m_o = m [(1 - v^2/c^2)^{1/2}]$$

Thus, if $v = 0$ (particle at rest) then we have:

$$m_o = m(1)^{1/2} = m$$

so the mass and rest mass are identical.

Now, the relativistic mass is then:

$m = m_0 / [(1 - v^2/c^2)^{1/2}]$

And again, if $v = 0$ then $m_0 = m$

But what if $v = c$? (Object moving at the speed of light?)

Then:

$m = m_0 / [(1 - c^2/c^2)^{1/2}] = m_0 / [1 - 1]^{1/2} = m_0 / 0 = \infty$

In other words, m would be infinite! This is another way of saying that to try to achieve the velocity of light one would have to overcome infinite inertia! In other words, it can't be done...not for a material object.

From this, we can also see the relativistic momentum must be:

$p = mu = m = m_0 u / [(1 - c^2/c^2)^{1/2}]$

This approaches the classical value ($p = mu$) as $u \rightarrow 0$

Newton's 2nd law in the relativistic format is simply:

$F = ma = m (du/dt) = d/dt[m_0 u / [(1 - v^2/c^2)^{1/2}]$

The relativistic energy is found by taking the integral of : $(dp/du) u\, du$

$\rightarrow \int_0^u u\, dp$

from 0 to u and obtaining:

$W = mc^2/ [(1 - u^2/c^2)^{1/2}] - mc^2$

And by the work -energy theorem:

$W = K(f) - K(i)$

where $K(i)$ is just the initial rest energy, or $m_o c^2$

Then $W = m_o c^2/ [(1 - u^2/c^2)^{1/2}] - m_o c^2 =$

(total energy - rest energy)

A variation of the above entails finding the work done (W') between two points x_1 and x_2 with velocity v_2 at x_2 and time t_2, and velocity v_1 at x_1 and time t_1.

Then we may write:

$W' = \int_{x_1}{}^{x_2} (dp/dt)\, dx = \int_{t_1}{}^{t_2} (dp/dt)(dx/dt)\, dt$

$= \int_{t_1}{}^{t_2} v (dp/dt)\, dt$

$= \int_{v_1}{}^{v_2} v\, dp = \int_{v_1}{}^{v_2} v (dp/dv)\, dv$

Note in the above we used the following facts:

(i) $v = dx/dt$ and (ii) $dp = (dp/dv)\, dv$

Here p is given as a function of v such that:

$W' = \int_{v_1}{}^{v_2} v (d/dv) [m_o v / [(1 - v^2/c^2)^{1/2}]\, dv$

On integrating the preceding equation by parts we find:

$$W' = m_o c^2 / [(1 - v_2{}^2/c^2)^{1/2}] - m_o c^2 / [(1 - v_1{}^2/c^2)^{1/2}]$$

This equation immediately shows that the effect of the work done is to produce a change in the quantity:

$$E' = m_o c^2 / [(1 - v^2/c^2)^{1/2}]$$

Note how this is different from the classical kinetic energy equation:

$$E = 1/2 \ (m v^2)$$

In particular E' does not become 0 when $v = 0$ (Instead it reduces to:

$$E' = m_o c^2$$

Hence, if we desire a quantity which correspond as closely as possible to classical KE we need to define:

$$E_k = m_o c^2 / [(1 - v^2/c^2)^{1/2}] \ - \ m_o c^2$$

If indeed this is a correct relativistic generalization of kinetic energy it must reduce to approximately:

$$1/2 \ m v^2 \quad when \ v << c.$$

This can easily be shown by expanding the binomial $(1 - v^2/c^2)^{1/2}$ in the last eqn. using the binomial theorem, i.e.

29

$$(1 - v^2/c^2)^{-1/2} = 1 + v^2/2c^2 + (3v^4/8c^4) + \ldots\ldots$$

Worked Problem (2):

Apply the basic mass-energy equation, $E = (\Delta m)c^2$, to the case of nuclear fusion below.

$$_1H^2 + _1H^2 \rightarrow _2He^3 + _2He^3 + _0n^1$$

which actually occurs in the Sun.

Solution:

We now arrange the masses (in atomic mass units) on each side:

$$2.015\ u + 2.015\ u \rightarrow 3.017\ u + 1.009\ u$$

or:

$$4.030\ u \rightarrow 4.026\ u$$

Note *the right side is less than the left* by an amount equal to the mass defect or:

$$\Delta m = 4.030\ u - 4.026\ u = 0.004\ u$$

To get the energy E:

$$E = (0.004\ u)(931\ MeV/u) = 3.7\ MeV$$

where 931 MeV/u is the conversion factor incorporating c^2

To transfer to Joules:

3.7 MeV = 3.7 MeV x (1.6 x 10⁻¹³ J/MeV)= 6.0 x 10⁻¹³ J

End of Chapter Problems:

1-Given that $x' = 1/a \, (x - vt)$ and $t' = 1/a \, (t - vx/c^2)$, derive similar equations for x and t in terms of x' and t'. (Recall: $1/a = (1 - v^2/c^2)^{1/2}$)

2- An event in space-time occurs at $x' = 60$ m, $t = 8$ x 10^{-8} s, in a frame S' ($y' = 0$, $z' = 0$). The frame S' has a velocity of 0.6c along the x-direction with respect to a frame S. The origins O and O' coincide at time $t = t' = 0$. Find the space-time coordinates of the event in S.

3- Suppose an astronaut is traveling at 0.9c in a space ship with respect to the Earth. How long a time interval will his clock indicate when the Earth has revolved once around the Sun? (Take the duration of one standard revolution of Earth around the Sun to be 365 ¼ days.)

4-The period T of a pendulum is measured to be T= 3.0 s in the inertial frame of reference of the pendulum. What is the period of the pendulum when measured by an observer moving at a speed of 0.95c with respect to the pendulum?

5. In the Michelson-Morley experiment, the length L of each arm of the interferometer was 11 meters. Sodium light of wavelength 5.9 x 10⁻⁷ m (590 nm) was used. The experiment would have revealed any fringe shift > 0.005 fringe.

What upper limit does this place on the Earth's velocity through the supposed Ether?

6- Using Fig.1, assume the time of travel to the right is: t(r) = L/(c - v) and the time of travel to the **left** is t(L) = L/(c + v).

a) Find the "total time of travel" by adding both left and right contributions.

b) Find the time consumed for "a half-trip".

c) Find the time consumed for a round trip.

d) Add the two "half trips" and what do you obtain?

e) Why does this not agree with the value obtained for (a)?

f) Work out what these times consumed would be for a trip to Proxima Centauri, using an ion –powered craft able to travel at v = 0.1c.

7-With what speed would a clock have to be moving to run at a rate that is one half the rate of a clock at rest?

8-An atomic clock is placed on a Jumbo Jet. The clock measures a time interval of 3600 s when the jet is moving at v = 300 m/s. What corresponding time would an identical clock left on the ground measure? (Hint: whenever v << c (e.g. v/c << 1), note that we have $1 + v^2/2c^2$ and not $[1 - v^2/c^2]^{1/2}$)

9-A muon formed high in the Earth's atmosphere

travels at v = 0.99c for a distance of 4.6 km before it decays into an electron, a neutrino and an anti-neutrino.

a) How long does the muon survive as measured in its rest frame?

b) How far does the muon travel as measured in its frame?

10- The average lifetime of a pi meson in its own frame of reference is 2.6×10^{-8} s. If the meson moves with v = 0.95c, what is its mean lifetime as measured by an observer on Earth?

11-Assume two astronauts are traveling at v = 0.95c on a journey to Alpha Centauri. We on Earth would say that it takes 4.2 / 0.95c = 4.4 years to reach the system 4.2 light years distant. But the astronauts dispute this.

(a) How much time passes on the astronauts' clocks?

(b) What is the distance to Alpha Centauri as measured by the astronauts? (Hint: this is an exact analog of the muon path length problem (#3) from the previous problem set)

12-According to Hubble's law, the distant galaxies are receding from us at speeds proportional to their distances, d, e.g. v = Hd. (Where $H = 2.26 \times 10^{-18}$ s^{-1}, currently).

a) How far away would a galaxy be in light years whose velocity relative to the Earth is c?

b) Would this galaxy be observable from Earth? (Take 9.5×10^{15} m = 1 LY)

13-A flash of light emitted at position x_1 on the x-axis is absorbed at position $x_1 + \ell$. In a reference frame moving with velocity $v = \beta c$ along the axis, what is the spatial separation ℓ' between the point of emission and point of absorption of the light? ($\beta = 1/ [1 - v^2/c^2]^{1/2}$)

How much time elapses between emission and absorption of the light?

14-A galaxy in Hydra emits light with a red shift corresponding to a recessional velocity of 6×10^4 km/s.

a) What is its distance according to Hubble's law?

b) What is the value of z?

c) Assume this galaxy passed Earth T years ago and has moved with constant velocity ever since, what is the value of T?

15-Some observations reported on the quasar 3C-9 suggest that when it emitted the light that just reached Earth it was receding at a velocity of 0.8c. One of the lines identified in its spectrum has a wavelength of 1200 Å (angstroms) when emitted from a *stationary source.*

a) At what wavelength must this spectral line have appeared in the observed spectrum of the quasar?

b) What is its red shift, z?

c) Find its *corrected velocity*, v.

16-A spaceship of mass 10^8 kg is to be accelerated to 0.6c using a matter-antimatter mix engine.

(a) How much energy does this require?

(b) How many kilograms of matter and antimatter will it take to provide this much energy?

17-Consider the decay:

$_{24}$ Cr ^{55} → $_{25}$ Mn^{55} + e-

The Cr 55 nucleus has a mass of 54.9279 u and the Mn 55 nucleus has a mass of 54.9244u.

(a) Calculate the mass difference between the two nuclei.

(b) What is the maximum kinetic energy of the emitted electrons?

18- Find the energy required to remove a simple proton from $_{19}$ K ^{41}.

19-Find the speed and mass of an electron whose kinetic energy is 50 MeV.

20-A rocket ship is to be accelerated to a speed of 0.5c. If propulsion is to be by using nuclear fuel, what fraction of the initial rest mass of the ship would have to be converted into kinetic energy to attain the desired speed? What time dilation results if the speed is v = 0.5c?

Would this be sufficient to allow one generation of humans to reach the star Proxima Centauri (4.2 light years distant)?

21-A Calcium line in the spectrum of α Centauri has a wavelength of 3968.20 Å. The same line in the solar spectrum has a measured wavelength of 3968.49 Å.

Find the *radial velocity* of α Centauri relative to the solar system. Is it approaching or receding?

22-A rocket ship of length 100m travels at v/c = 0.6. It carries a radio receiver in its nose. A radio pulse is emitted from a stationary space station just as the ship passes by.

a) How far from the space station is the nose of the rocket at the instant the radio signal arrives at the nose?

b) By space station time, what is the time interval between the arrival of this signal and its emission from the station?

c) What is the time interval determined from measurements in the rocket ship's rest frame?

Chapter II. Basic Atomic Physics

2.1. The Rutherford Model of the Atom.

What may be called the first foray into basic atomic physics by which further theory could be built upon, commenced with the Geiger and Marsden experiment – first suggested by Lord Rutherford in 1909. The basic setup is depicted in the rough sketch below:

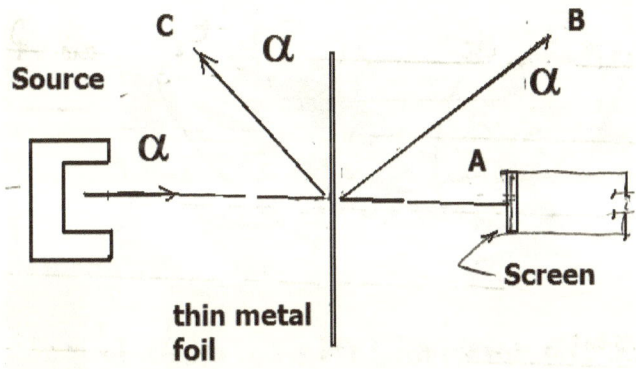

Fig..2. 1: Layout of the Geiger-Marsden Experiment

From the Rutherford experiment design, Geiger and Marsden made use of a source of alpha particles to bombard a thin metal foil, on the other side of which was a detecting zinc sulfide screen. They found that while most alpha particles arrived at A, in the direction shown, a few also scattered to positions at B and C which could be detected when the screen at A was moved to the other positions. The nature of the scattering and deflections (especially some alpha particles at very large angles) was such that there had to be a highly concentrated charge or "nucleus" at the

37

center of the atom. Since the alpha particles are relatively massive (at about 4.002 amu each) the deflections at wide angles meant nearly all the atomic mass was concentrated in the center of the atom and electrons were in the distant outer regions.

Rutherford thereby proposed a model of the atom in which nearly all the mass was concentrated in a very small nucleus while the electrons were scattered at some distance away. This is depicted below in Fig. 2.

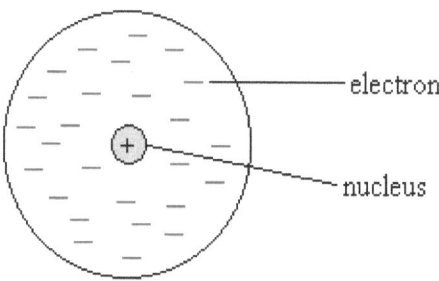

Fig. 2.2: The Rutherford Model of the atom.

The key consequence was that the Rutherford experiment, carried out by Geiger and Marsden, showed that the "pudding pie" model of J.J. Thomson was incorrect. If Thomson's model was correct, then the expected deflection could be no larger than 0.0001 radians or less than a degree. Since the observed deflections were in some cases more than 100 degrees, it failed the experimental test.

Despite this success, Rutherford's model still hadn't won the day. It was largely accepted because it could quantitatively alpha-scattering by thin foils. His model could not: 1) explain line spectra in atoms, including both absorption and emission lines, 2) account for the stability of atoms and could only account for half the nuclear mass.

2.2. *The Bohr Model of the Atom*.

The Bohr Model of the atom, proposed by Neils Bohr, directly challenged the Rutherford model by showing how the observed emission and absorption lines of spectra could be explained. At the heart of Bohr's model was simplicity, with the hydrogen atom – for example – configured to a miniature solar system with the nucleus at the center and the electron in orbit around it.

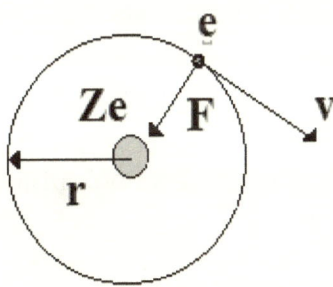

Fig. 2.3: The Bohr Model of the atom.

From the diagram the electron (e) orbits at a radius r from the central nucleus of charge Ze. As with the planets, a centripetal (inner directed force) F acts toward the center.

Bohr's major concept was to quantize the electron orbits. He proceeded by first quantizing the angular momentum of the orbit:

$$m\,vr = nh/\,2\pi = n\,\hbar$$

where $\hbar = h/2\pi$ is the Planck constant divided by 2π.

The Planck constant, first proposed by Max Planck, is:

$h = 6.626069 \times 10^{-34}$ J-s

Then the value of $\hbar = 1.0546 \times 10^{-34}$ J-s

Next: both sides are squared:

$(m\,vr)^2 = (n\,\hbar)^2$

So: $m^2 v^2 r^2 = n^2 \hbar^2$

And: $v^2 = n^2 \hbar^2 / m^2 r^2$

Bohr looked at the total energy of the Hydrogen atom in terms of it kinetic (K) and potential (V) contributions, so:

$E = K + V = \tfrac{1}{2} m v^2 - k e^2 / r$

$E = K + V = k e^2 / 2r - k e^2 / r = - k e^2 / 2r$

(Since $\tfrac{1}{2} m v^2 = k e^2 / 2r$)

Now solve for r (actually the *quantized* r_n):

$r_n = [\, n^2 \hbar^2 / m^2 v^2 \,]^{1/2}$

But from the kinetic energy equivalence:

$v^2 = k e^2 / mr = n^2 \hbar^2 m^2 r^2$

$\therefore \quad r_n = [n^2 \hbar^2 / m k e^2]$

The Bohr radius is just the value when the principal quantum number n = 1, so :

$r_0 = [\hbar^2 / m k e^2] = 0.0529$ nm $= 5.2917 \times 10^{-11}$ m

This is just the most probable radius, i.e. distance between proton and electron, in the hydrogen ground state. To obtain the quantized energy (E_n) substitute the value for r_n into the total energy equation:

$E = - k e^2 / 2r = - k e^2 / 2[n^2 \hbar^2 / m k e^2]$

$E = - m k^2 e^4 / 2n^2 \hbar^2 = - m k^2 e^4 / 2 \hbar^2 \ [1/ n^2] = -13.6/ n^2$

Where the last quantity is in *eV*, or *electron volts*. Here the n refers to the energy level, ground state is n = 1, so can allow the computation of energy for a given level. Or, the energy for a photon emitted from an atom when an electron makes a transition – say from n = 2 to n = 3.

An important point is that the **quantized angular momentum postulate** (m vr = n $\hbar$) restricts the possible circular orbits to defines sizes according to the quantized radii (r_n etc.).

Thus the normal state of the atom, say hydrogen, will be that for which it has the *least energy* so occupying the ground state – corresponding in the case of hydrogen to the Bohr radius. Such a situation is shown below:

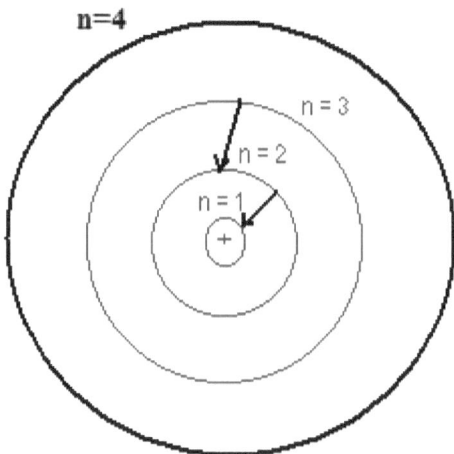

Fig. 2.4: A few energy transitions made in Hydrogen

Some specific transitions for different spectral series are shown below:

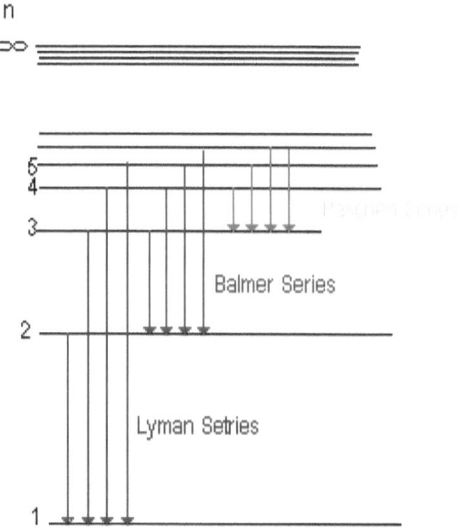

Fig. 2.5. Some Energy transitions in the Hydrogen Bohr atom

Emission occurs when an electron in the atom, say hydrogen, makes a transition from a *higher to a lower* energy level. A photon results with a defined energy E = hf = h (c/ λ). Consider a transition from the n = 2 to the n = 1 level, as depicted in the lower right of Fig. 4 and *in the first line of the Lyman series* of Fig. 5.

The energy at the n= 2 level is:

E(n=2) = - 13.6/ n^2 = - 13.6/ $(2)^2$ = - 13.6/4 (eV)

Now, 1 eV = 1.6 x 10^{-19} J so:

E(n=2) = - 13.6/4 (eV) = -(3.4) x 1.6 x 10^{-19} J =

-5.4 x 10^{-19} J

The **n= 1 level** has energy:

E(n=1) = - 13.6/ n^2 = - 13.6/ $(1)^2$ = - 13.6 (eV)

E(n=1) = -(13.6) x 1.6 x 10^{-19} J = -21.8 x 10^{-19} J

Then the energy difference is:

E2 – E1 = [- 5.4 – (-21.8)] x 10^{-19} J = 16.4 x 10^{-19} J

From this, the wavelength of the photon emitted can be found. Since E = hf = h (c/ λ):

λ = hc/ (E2 – E1)

λ =
6.626069 x 10^{-34} J-s)(3 x 10^{8} m/s)/ 16.4 x 10^{-19} J

$\lambda = $ 1.21 x 10 $^{-7}$ m

The frequency can be found from:

f = (c/ λ) = (3 x 10 ^{8} m/s) / 1.21 x 10 $^{-7}$ m = 2.47 x 10 ^{15} Hz

Inquiry- Insight problems:

(1)Using Fig. 5 as a basis, compute the energies and wavelengths of the photons emitted when the electron in the hydrogen atom makes the 1st, 2nd and 3rd Balmer transitions.

(2) Show that the energy E of a photon and its wavelength λ are related by:

$E\lambda = 1.99 \times 10^{-16}$ J nm

3) The diagram below represents part of the emission spectrum of atomic hydrogen. It contains a series of lines, and the wavelengths of some (in nm) are marked. There are no lines in the series less than 91.2 nm

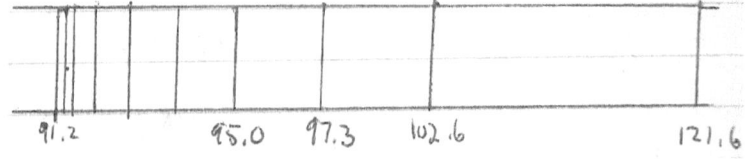

91.2   95.0   97.3   102.6   121.6

(a) In which region of the E-M spectrum would these lines occur?

(b) Using the relation between *E and λ* in (1) find the photon energies equivalent all the wavelengths marked.

2.3. The Photo-Electric Effect:

The famous photo-electric effect is important because of how it highlights the particulate nature of quanta. While electron diffraction enabled the hypothesis of matter waves or de Broglie waves, the photo-electric effect reinforced the nature of light as **photons**.

The effect was first observed by Heinrich Hertz in 1887, but it was left for Einstein to explain (and for which he won the Nobel Prize) in 1905. The effect at the time, was most directly observed when a + charged zinc plate (in a Braun type electroscope) was exposed to x-rays or ultraviolet radiation which caused an increased deflection of the electroscope leaf. Conversely, a negatively charged plate exposed to the same high frequency radiation caused a decreased deflection showing a loss of potential. Hertz demonstrated the effect using an apparatus such as shown in Fig. 6.

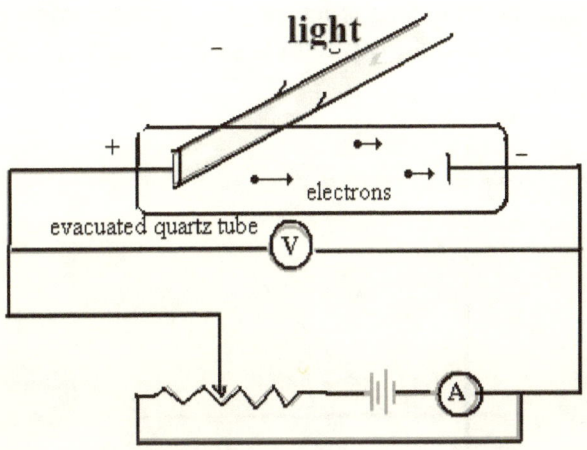

Fig. 2.6: Apparatus to investigate the photo-electric effect

Here, an evacuated tube contained two electrodes connected to an external circuit with the anode being the metal plate on which the radiation was incident. The photo-electrons emerging from the surface thus had sufficient kinetic energy to reach the cathode despite its negative potential. These electrons formed the current (photo-current) measured by the ammeter.

To measure the maximum kinetic energy of the photo-electrons one applies a retarding voltage V, gradually increasing it until the most energetic photo-electrons are stopped so the photo-current becomes zero.

At this point: $eV_s = K_{max} = \frac{1}{2} mv_{max}^2$

Thus, the maximum kinetic energy of the electrons can be obtained if V_s is known. If K_{max} is then plotted against the frequency of the incident radiation (for different tests) a graph such as that shown in Fig. 7 is obtained.

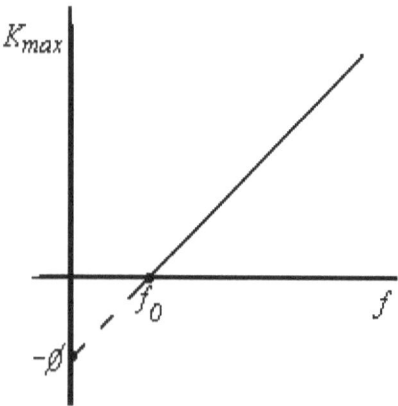

Fig. 2.7: Plot of K max vs. f to get the work function

The results of such experiments showed:

i)The number of photo-electrons emitted is proportional to the intensity of the incident radiation

ii) The photo-electrons are emitted within an energy range; $0 < K < K_{max}$ corresponding to a range of frequencies: $f_o \leq f \leq f'$.

Hence, there exists some frequency (f_o) defined as the threshold frequency, *below which no electrons are emitted.*

From the graph originating in such experiments, it is therefore possible to write:

$$\tfrac{1}{2} m v_{max}^2 = hf - \phi$$

Where ϕ is the "work function". It follows from this that one can also get the following graph in terms of the *stopping potential* V_s :

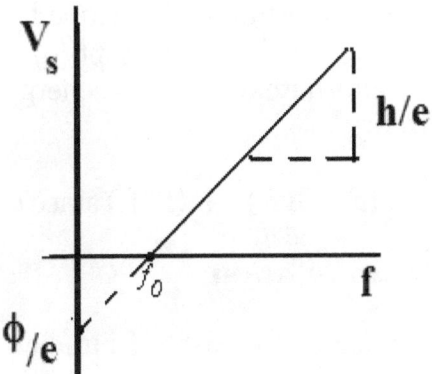

Fig. 2.8: Graph for the stopping potential

This is known as the empirical graph based on actual measurements. Recall from the theory:

$$\tfrac{1}{2} m v_{max}^2 = hf - \phi$$

And, from experiment: $eV_s = \tfrac{1}{2} m v_{max}^2$

Therefore, combining the two:

$$eV_s = hf - \phi$$

or:

$$V_s = hf/e - \phi/e = (h/e) f - \phi/e$$

Where (h/e) is then *the slope* which can be computed based on the two known quantities ($h = 6.626 \times 10^{-34}$ Js) and $e = 1.6 \times 10^{-19}$ C. This yields:

$(h/e) =$
$(6.62 \times 10^{-34} \text{ Js})/ 1.6 \times 10^{-19} \text{ C} = 4.13 \times 10^{-15}$ Js/C

Hence, in such experiments the slope h/e will always remain the same but the y-intercept (ϕ/e) will change. Note that an alternate form of the energy relationship can be written:

$$\tfrac{1}{2} m v_{max}^2 = hf - (h f_0) = h (f - f_0) \text{ since } \phi = h f_0$$

Einstein's explanation:

A beam of radiation consists of bundles of energy of size hf called "photons". When such photons collide with electrons at or on a metal surface, they transfer an energy hf. The electrons on the metal surface either get all of this energy or none at all. In leaving the surface, electrons lose an amount of energy ϕ which is the work function of the surface. The maximum energy with which an electron can emerge is:

(Energy gained from work function) – (work function)

Worked Problem:

Sodium has a work function of 2.0 eV. Calculate the maximum energy and speed of the emitted electrons when sodium is illuminated by radiation of λ = 150 nm. What is the lowest frequency of radiation for which electrons are emitted?

Solution:

The work function: $\phi = 2$ eV $= 2(1.6 \times 10^{-19}$ J$)$

$\phi = 3.2 \times 10^{-19}$ J

The incident energy $E = hf = hc/\lambda$

$hc/\lambda = (6.62 \times 10^{-34}$ Js$)$ $(3 \times 10^8$ ms$^{-1})/$ $(150 \times 10^{-9}$ m$)$

$hc/\lambda = 13.2 \times 10^{-19}$ J

Therefore:

$K_{max} = hf - \phi = [13.2 \times 10^{-19}$ J $- 3.2 \times 10^{-19}$ J$]$

$K_{max} = 10^{-18}$ J

The velocity $v = \sqrt{(2\,K_{max}/m)} =$

$[(2 \times 10^{-18} \text{ J}) /(9.1 \times 10^{-31} \text{kg})]^{1/2} = 1.5 \times 10^{6} \text{ ms}^{-1}$

Threshold frequency $f_o = \phi / h$

$\phi / h = (3.2 \times 10^{-19} \text{ J})/ (6.62 \times 10^{-34} \text{ Js})$

Therefore:

$f_o = 4.8 \times 10^{14} \text{ Hz}$

2.4. Radioactive Activity and Decay:

Radioactivity basically occurs in three forms, as determined by the particles: alpha-radiation (from alpha particles or Helium nuclei), beta radiation, from beta particles or electrons, and gamma radiation, from gamma particles or very high energy photons

The diagram of Fig. 9, using a simple experiment, graphically shows the differences in the radiations with respect to an applied magnetic field.

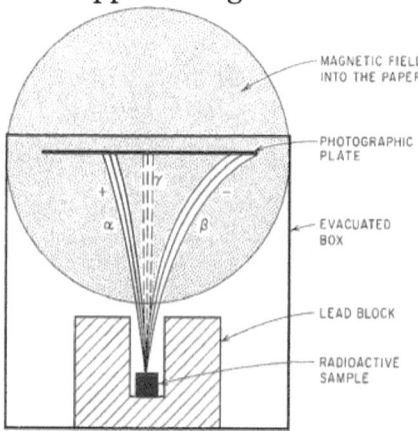

Fig. 2.9: Different radiations in a magnetic field

If one holds the thumb of one's RIGHT hand *into the image* (to represent the B-field direction) then the electrons (beta particles) will display a direction coincident with the curving fingers of the right hand. That is, clockwise. Since the alpha particles are *positively charged* (, e.g. He++, as opposed to the negatively charged beta particles, e.g. e-) they will go in the opposite direction.

The gamma rays, meanwhile, suffer no deflection in the field because they have zero charge, being photons of light. In terms of penetration power these additional differences apply:

1) Alpha (α) particles are absorbed by a few cm of air, or by an aluminum foil only 0.006 cm thick

2) Beta (β) particles - while having less ionizing power than alpha particles (because of much lower mass) have 100 times more penetrating power. A sheet of aluminum at least 3mm thick is needed to absorb them.

3) Gamma rays produce little ionization since they have no electric charge but can pass through a block of iron a foot thick.

The Activity of a radioactive source.

We define the activity of a radioactive source as:

$$A = dN/dt = -\lambda N$$

Where λ is the *decay constant*. The negative sign appended to the equation indicates that the amount N is decreasing with time t.

The units are defined as follows:

$A \, [Bq] = dN/dt \, [s^{-1}] = -\lambda N [s^{-1}]$

Where Bq is Becquerels.

The decay curve is obtained from the fundamental law of radioactive decay, based on some original number of atoms N_o decaying with an activity λ over time t:

$N = N_o \exp(-\lambda t)$

Then: $\left| d \, N/dt \right| = N_o \, \lambda \exp(-\lambda t) = R$

Where R is the decay rate, i.e. $R = R_o \exp(-\lambda t)$

And: $R_o = N_o \, \lambda$ is the decay rate at time t = 0.

The half-life is the time for half of the original (N_o) atoms to disintegrate or when the point is reached such that:

$N \rightarrow N_o /2$ or $R \rightarrow R_o /2$

Then:

$N_o /2 = N_o \exp(-\lambda T_{1/2})$

Where $T_{1/2}$ is specifically substituted for t.

After dividing N_0 into both sides and taking natural logarithms we get:

$$\lambda T_{1/2} = ln\ 2 = 0.693$$

Or:

$$T_{1/2} = ln\ 2/\lambda = 0.693/\lambda$$

Using this basis any sample or fossil with even a minuscule amount of radioactive material can be dated. All we need know is that over the period defined as $T_{1/2}$ half of the number of the remaining atoms decay and the activity is in Becquerels (Bq). Thus, if $T_{1/2} = 15{,}000$ yrs. for $\lambda = 200$ Bq *then if $\lambda = 50$ Bq now the sample is 45,000 years old.*

A graphical depiction of generic radionuclide decay is shown below in Fig. 10.

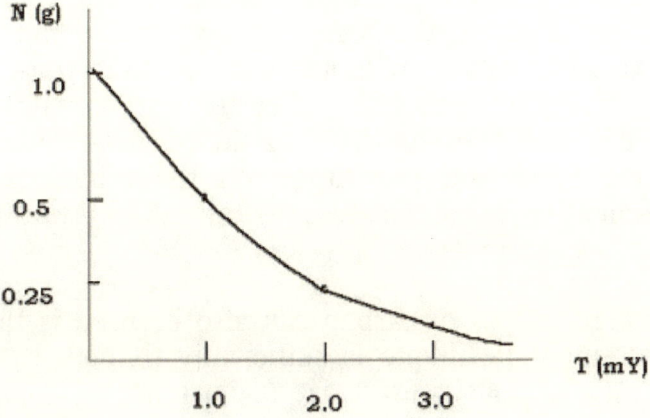

Fig. 2.10: Radioactive Decay for a radionuclide with time T in millions of years (mY)

Here the vertical axis shows a relative scale for the amount or mass of some, unnamed decaying isotope which commences decay at some initial specified value, e.g. 1 gram then decreases to half that original amount in one half life. It is easy to see from the graph shown that the half-life of this nuclide is about 1 million years, so: $T^{1/2} = 10^6$ yrs.

The problem of dating radioactive fossils or other specimens (e.g. the cloth of Turin) is really a problem of finding a radioactive isotope that is most appropriate, or one that enables the maximum accuracy for the time scale desired. Hence, for ancient fossils one would look for isotopes that have half-lives in the millions rather than thousands of years. Failing that one would wish to have available some kind of correction method, say to correct for extraneous effects such as the atmosphere might ipose on samples.

In many ordinary fossil dating applications, potassium-argon methods are employed, based on the relative compositions of Potassium -40 to Argon-40 gas. Typically when rocks or other items are tested the sample is split between the Potassium-40 content on the one hand and the Argon on the other. The instrument of choice to assess the ratio: K40/Ar 40 is the mass spectrometer.

Exotic isotopes of carbon can also be used is the measurement technique is sufficiently refined. In a recent use of the isotope $\delta\ ^{13}$ C, for instance, evidence has been found for the existence of life on Earth at least 3,850 million years ago. For this purpose, quartz (zircon, zirconium) crystals have been found to be of use since they may harbor small amounts of thorium and uranium at the level of parts per billion.

For dating samples in the millions of years, particularly for igneous rocks and samples embedded within, isotopes of lead and strontium may be used – being the 'daughters' from millions of years of radioactive decay.

Practical:

A Geiger-Muller tube measures the background count at a given place to be 20/min. Over a period of time, the readings shown in the table below are obtained after an unknown source is placed at the location.

Time hrs	0	6	8	10.5	20
Count/min	120	70	60	50	30

The purpose of the practical is to obtain the corrected counts, determine the half -Life ($T\frac{1}{2}$) of the unknown source, and find the decay constant.

Procedure:

First, we obtain *the corrected counts* by subtracting the background count of 20/m from each of the values above, to obtain:

Correct/cpm	100	50	40	30	10

From inspection, since the activity drops by half (from 100 to 50 cpm) in 6 hours, 6 hrs is the half life.

Hence: $T\frac{1}{2}$ = 6 h = 21600 s

The decay constant λ is found from:

$$\lambda = \ln 2/ (T\tfrac{1}{2}) = 0.693/ (21600s) = 3.2 \times 10^{-5} /s$$

This can also be verified by sketching a plot of the half life (($T\tfrac{1}{2}$)) on the vertical axis against the time in hours - to obtain the radioactive decay curve. Such a decay curve which would represent the half-life for this problem is shown below:

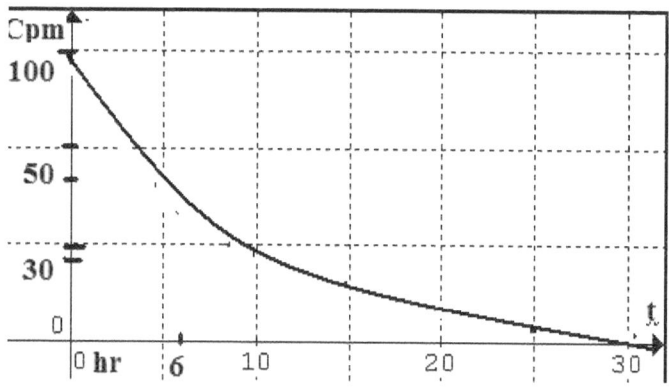

Fig. 2.11: Decay curve result from practical

2.5. Electron Spin & Complex Electron Atoms

In describing more complex atoms we need three different quantum numbers: n (principal), ℓ (angular momentum), and m ℓ (magnetic). We now examine how the fourth quantum number (m s) or electron spin came about.

By the early 1920s, quantum mechanics (next chapter) had developed to the point that theorists realized an electron magnetic moment ought to exist.

In any external magnetic field **B**, the magnetic moment ought to experience a force similar to what a magnetic compass needle experiences in Earth's magnetic field. According to quantum mechanics, the assumed values should be quantized, so the magnetic moment can assume only certain values, given by:

$2 \ell + 1$

Thus, the orientation number depends on the second or azimuthal quantum number, ℓ. The total angular momentum (L) would therefore be:

$L = \hbar \ [\ell \, (\ell + 1)]^{1/2}$

In 1922, the German physicists Otto Stern and Walther Gerlach passed a beam of electrons through a non-uniform magnetic field **B** as shown in the top sketch of the diagram (C denotes collimator and D, detector plate). A non-uniform field meant the field was stronger on one side of the beam than the other. As predicted from theory, the force on the magnetic moment of the electrons is such that the field ought to deflect the beam according to the orientation of its moment.

The field should therefore split the beam into $2 \ell + 1$ parts (according to $2 \ell + 1$ orientations). Stern and Gerlach found their beam of hot silver atoms split into two parts. At first this appeared surprising and at odds with theory but later work showed the conflict could be resolved if the electrons going into the opposing (up and down) beams, had their own spins, or intrinsic angular momenta.

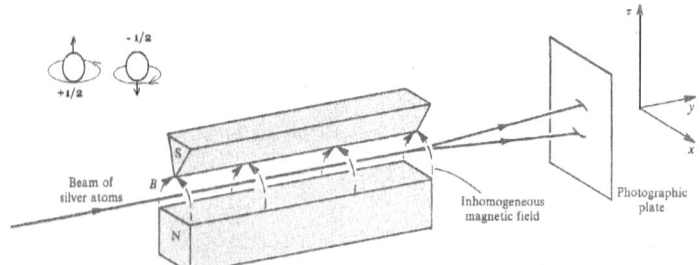

Fig. 2.12. The Stern-Gerlach Experiment

The upper beam features electrons with "*spin up*" angular momentum of (+1/2) while the lower beam would feature electrons with "*spin down*" angular momentum of (-1/2) . These are shown at the lower section of the diagram. Note that in each case, spin up or down, the orientation accords with the direction of rotation: clockwise or anti-clockwise.

The refining experiment was actually done by Phipps and Taylor using a beam of heated hydrogen atoms. They began by defining the *magnetic dipole moment* such that:

$\mu_{\ell z} = - g(\ell) \, u_B \, m_\ell$

where, as before, $m_\ell = - \ell, - \ell +1, 0, ...+ \ell$

and u_B is the *Bohr Magneton*: $u_B = eh/4\pi \, m$

where m is the mass of the electron.

Meanwhile, $g(\ell)$ is known as the "*orbital g-factor*".

Phipps-Taylor recognized that $\ell = 0$ for the electrons coming off and entering the magnetic field.

58

Therefore, $m_\ell = 0$. Since this was so, then:

$\mu_{\ell z} = 0$

Phipps -Taylor assumed the beam would be **unaffected** or not split at all. Yet, they then observed it split into two symmetrical components. Given the earlier Stern-Gerlach experiment, plus their own, they therefore had to expect the electron had its own magnetic dipole moment.

They assumed this to be a spin magnetic moment μ_s, due to the electron having an intrinsic angular momentum S analogous to L (angular momentum) so that:

$S = [s(s + 1)]^{1/2} \hbar$ and $S_z = m_s (\hbar)$

Where S_z is the z-component of angular momentum

Then m_s is the electron spin or spin quantum number, 1/2 or - 1/2,

Then the electron magnetic dipole moment would be:

$\mu_s = - g(s) u_B / \hbar [S]$

To nail down the basis quantitatively, Phipps-Taylor knew that the net force felt by the electrons traversing the field would be:

$F_z = - (dB_z/dz) u_B g(s) m_s$

where m_s is the putative spin quantum number.

Since they knew the Bohr magneton :
$u_B = 9.27 \times 10^{-24}$ J/T

And dB/dz could be measured (e.g. the difference in the B-field over the collimation width dz).

Then : $g(s) \, m_s = F_z / - (dB_z/dz) \, u_B = \pm 1$

Since $g(s) = 2$, then by deduction: $m_s = \pm 1/2$

Phipps and Taylor discovered that the splitting effectively showed two possible values for energy:

$\Delta(E) = - \mu_s B = \pm g(s) \, u_B \, B/2.$

Let's compute S for the value $m_s = 1/2$.

$S = [s(s+1)]^{1/2} \hbar = [1/2(1/2 + 1)^{1/2} \hbar =$

$[1/2 \, (3/2)]^{1/2} \hbar = [3/4]^{1/2} \hbar = (\sqrt{3}/2) \hbar$

This result can be shown on a spin diagram, e.g.

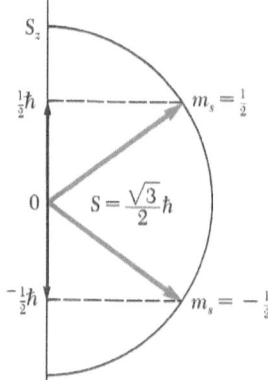

Fig. 2.12. Illustrating spin quantization

Note how the magnitudes for S_z are first laid out on the vertical axis with lines drawn to a semi-circle, then the arrows are drawn to intersect with the S_z values of $\pm \frac{1}{2}$.

Worked Problem:

Write out the electron configuration for oxygen (O16), then write out the values for the set of quantum numbers : n, ℓ, m_ℓ, m_s for each of the electrons in O16

Solution:

Finding the electron configuration means one must use the *Pauli Exclusion Principle* to make sure the electrons are distributed so that *no two electrons have the same set of quantum numbers*. This means we need to have 2 in the 1s shell, 2 in the 2s shell, and 4 in the 2p shell, so:

1s(2) 2s(2) 2p(4)

For the 2 electrons in the 1s shell: n = 1, ℓ = 0, = 0 and $m_s = \frac{1}{2}$ (and (- $\frac{1}{2}$), for second)

For the 2 electrons in the 2s shell: n=2, ℓ =0, m_ℓ = 0 and $m_s = \frac{1}{2}$ (and (- $\frac{1}{2}$), for second)

For the 4 electrons in the 2p shell: Since ℓ =1 corresponds to p: n=2, ℓ =1, m_ℓ = +1, $m_s = + \frac{1}{2}$

Then: n=2, ℓ =1, m_ℓ = -1, $m_s = -\frac{1}{2}$

And: n=2, ℓ =1, m $_\ell$ = 0, m $_s$ = +½

Finally: n=2, ℓ =1, m $_\ell$ = 0, m $_s$ = -½

2.6. Space quantization and L-S Coupling:

We now venture a bit deeper into the hydrogen atom with a view to doing more elaborate problems, ultimately leading to tangling with what we call "*L-S Coupling*", or combining the orbital angular momentum, L, with spin angular momentum, S, to arrive at the J value.

In terms of an orbiting planet, for example, we can specify the classical orbital angular momentum as: L = mvr. This assumes a circular orbit of radius r, and an orbiting body of mass m, with velocity r.

What about at the atomic level? In his simplified model, Bohr visualized electrons behaving like tiny miniature planets and orbiting a central nucleus. For the simplest atom, hydrogen, this meant one electron circling a single central proton. Bohr specified the rule for orbital angular momentum - using the principal quantum number, n, as:

mvr = n ($\hbar$)

where again, h is Planck's constant (6.6254 x 10 $^{-34}$ J-s) and $\hbar$ the reduced Planck constant.(h/ 2π)

However, this model failed to correctly predict the orbital angular momentum for the hydrogen electron, yielding a value of 1 unit, which is wrong. In addition, if L = 0 one would find the electrons oscillating in a straight line *through the proton nucleus* - which is impossible! Thus, it had to be modified. The modifications were possible once one removed the "planetary model" (which is classical and deterministic) and turned to the wave model. In this case, the orbital angular momentum assumes certain specific values such that:

$L = [\ell (\ell + 1)]^{1/2} (\hbar)$ where: ℓ = 0, 1, 2, ..., n-1

When ℓ = 0 we find $L = \hbar$ =1. 054 x 10 $^{-34}$ J-s

The fact L can be zero and is acceptable discloses why classical mechanical models fail at the quantum level.

What if instead we have the angular momentum quantum number ℓ = 2?

Then: $L = [\ell (\ell + 1)]^{1/2} (\hbar) = [2(2 +1)]^{1/2} (\hbar) = \sqrt{6} (\hbar)$

And a set of allowed projections for L are obtained:

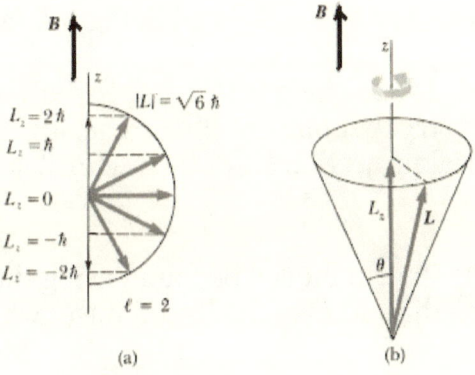

Fig. 2.13. Space quantization and quantum number, L

Of particular interest are the allowed values for the vertical component of the orbital angular momentum vector, which we designate: L_z. (L_z is the projection of L along the z-axis and has discrete values, e.g. 0, $h/2\pi$, h/π, $-h/2\pi$, $-h/\pi$ for the case given).

In general L_z is specified according to:

$$L(z) = m_\ell \hbar$$

where m_ℓ is the magnetic quantum number.

The key point here is that the *direction* of the orbital angular momentum quantum number L is quantized with respect to an external magnetic field. We call this "*space quantization*".

We emphasize that m_ℓ can range from $-\ell$ to ℓ, so for $\ell = 2$, we have:

$$m_\ell = +2, +1, 0, -1, -2$$

When we multiply each of these by ($\hbar$) we obtain the quantization of L_z as depicted in Fig. 1.6 (a).

It should also be obvious to anyone who's taken trigonometry that one can obtain the angle between the vertical projection L_z of the vector L, and L. (See 1.6 (b)).

The angle (θ) can indeed be found using the cosine relation (adjacent over the hypotenuse) which yields:

$$\cos(\theta) = L_z / L = m_\ell / [\ell (\ell + 1)]^{1/2}$$

Of course, it ought to be self-evident that we are obtaining allowed values for the angle, since obviously m $_\ell$ is going to range from -2 to +2)

Thus, for the problem we've considered (with ℓ = 2) we have the allowed cosines and angles:

m $_\ell$ = 0 so $\cos(\theta)$ = 0 so t = 90 deg

m $_\ell$ = 1 so $\cos(\theta)$ = 1/ $\sqrt{6}$ so θ = 65.9 deg

m $_\ell$ = -1 so $\cos(\theta)$ = -1/ $\sqrt{6}$ so θ = 114.1 deg

m $_\ell$ = 2 so $\cos(\theta)$ = 2/$\sqrt{6}$ so θ = 35. 2 deg

m $_\ell$ = -2 so $\cos(\theta)$ = -2/$\sqrt{6}$ so θ= 144.7 deg

Question: Say an electron in an atom (e.g. hydrogen) has zero orbital angular momentum (ℓ = 0) does that mean it has zero *total angular momentum*?

No, because in atomic physics we find that for every electron in a given atom we have to process two kinds of angular momentum: orbital (L) and spin (S). Even if the electron experiences no precession or torque it must still exhibit a total angular momentum.

Earlier, we saw for the total orbital angular momentum:

$L = [\ell (\ell + 1)]^{1/2} \hbar$

We also saw that each electron has a spin angular momentum: $S = [s(s + 1)]^{1/2} \hbar$

For which s assumes one or other of the electron spin quantum numbers, $m_s = +1/2$ or $-1/2$.

It can be shown that S is always: $[3/4]^{1/2}\ \hbar$

Now, we reckon in what we call the total angular momentum or J, such that:

$J = [j(j + 1)]^{1/2}\ \hbar$

and $j = \ell + s$ (note the common letters apply to different quantities than the capital ones!)

Thus, for an electron with zero orbital angular momentum (l=0) we have:

$j = \ell + s = 0 + 1/2$ or $\ell + s = 0 - 1/2$

so: $j = +1/2$ or $-1/2$

Then we have:

$J = [j(j + 1)]^{1/2}\ \hbar\ = [3/4]^{1/2}\ \hbar$

for either j (which readers can also verify)

We can also find (as we did with L for L_z, the projection of the total angular momentum quantum number on the z-axis (J_z) which will be:

$J(z) = m_J\ \hbar$

where $m_J = -j, -j+1, ...+j$

Readers with an intuitive grasp of vectors, or if they've

worked with vectors - say in high school or college physics, will quickly see that the name of the game is to obtain a vector sum such that:

V = V(1) + V(2)

In this case, L plays the role of V(1), and S plays the role of V(2)

Then we obtain for **J**:

J = L + S

As any physics student knows, the way to obtain the vector sum is via the law of cosines and this is demonstrated in the diagram below along with the computations. This is for the case:

L = 3

S = 1/2

J = 5/2

The angle (θ) can also be obtained as shown below:

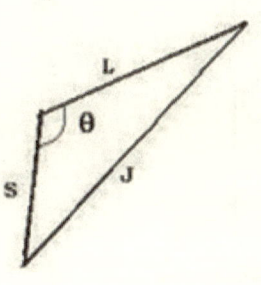

Angle between atomic vectors L and S is what we wish to determine

For which:

$$\theta := acos\left[\frac{J\cdot(J+1) - L\cdot(L+1) - S\cdot(S+1)}{2\cdot\left[\sqrt{L\cdot(L+1)}\cdot\sqrt{S\cdot(S+1)}\right]}\right]$$

$\theta = 2.300523983022$

Note that in a weak magnetic in which the atom is situated, the L-S coupled system depends on J, in other words this very angle between the vectors L and S. The orientation of the atom depends on m $_J$.

Worked Problem (2):   State which values of ℓ, s, and j would apply to the preceding diagram with L= 3, S= ½ and J= 5/2. , Indicate the relations on a diagram such that the assorted angular momentum vectors have the values identified.

Solution: From the diagram below, if L = 3 we get two possible values of ℓ.

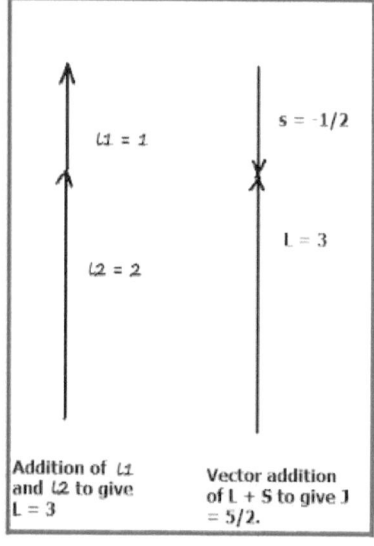

So $\ell_1 = 1$ and $\ell_2 = 2$ therefore: $\ell_1 + \ell_2 = 1 + 2 = 3$.

Meanwhile, S can be defined by only one value of s, or $s = \frac{1}{2}$

The possible j-values are:

$j = \ell + s = \ell_1 + s = 1 + \frac{1}{2} = 3/2$

$j = \ell_2 + s = 2 + \frac{1}{2} = 5/2$

$j = \ell - s = \ell_1 - s = 1 - \frac{1}{2} = \frac{1}{2}$

$j = \ell_2 - s = 2 - \frac{1}{2} = 3/2$

So in total: $j = \frac{1}{2}$, $3/2$ and $5/2$

Note that, conforming to j-selection rules, all the j's differ by *an integral amount*, though they are half-integral (e.g. $3/2$, $5/2$) themselves.

To obtain any J (total angular momentum) we need an L-S coupling vector that yields $J = 5/2$. Two possible L-S couplings are available: [L + S] and [L + S − 1] and it is the last that yields the appropriate result: $[3 + \frac{1}{2} - 1] = 5/2$

This means we need values such that $\ell_1 = 1$, $\ell_2 = 2$ and $s = \frac{1}{2}$ to make this work. (See Fig. 1.7 below and how the right side discloses the J result)

Worked Problem (3):

Enumerate all the possible values of j and the

quantum number m J for which $\ell = 3$ and $s = \frac{1}{2}$, but not connected at all to the original problem.

Solution:

$j = \ell + s = 3 + \frac{1}{2} = 7/2$

and:

m J $= -7/2, -5/2, -3/2, -\frac{1}{2}, \frac{1}{2}, 3/2, 5/2$ and $7/2$

meanwhile:

$j = \ell - s = 3 - \frac{1}{2} = 5/2$

So:

m J $= -5/2, -3/2, -\frac{1}{2}, \frac{1}{2}, 3/2, 5/2$

Note that these last m J quantum numbers would be the ones applicable to the original problem for which: $L = 3$, $S = \frac{1}{2}$, and $J = 5/2$.

Worked Problem (4):

Enumerate the possible values of j and m J for the states in which $\ell = 1$ and $s = 1/2$ and draw the associated vector diagrams.

Solution:

$j = \ell + s = 1 + \frac{1}{2} = 3/2$

The diagram for the vectors is shown below:

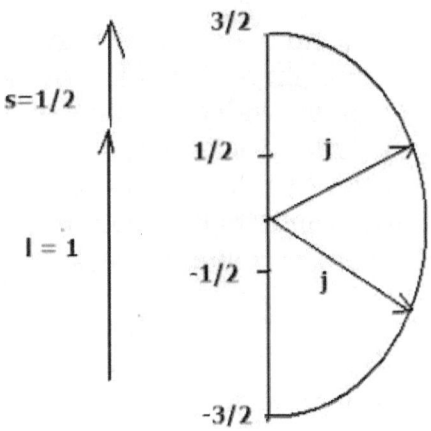

then: $m_J = -3/2, -\tfrac{1}{2}, \tfrac{1}{2}, 3/2$

and

$j = \ell + s = 1 + (-\tfrac{1}{2}) = 1/2$

so:

$m_J = -\tfrac{1}{2}, \tfrac{1}{2},$

2.7. *Applications to atomic configurations*:

We can apply the preceding rules to a particular atomic configuration, say: 2p 3d then go through a procedure of taking differences, assigning values etc. based on selection rules. For the configuration given, one particular energy level occurs for:

$s' = 1 \; (s' = \tfrac{1}{2} + \tfrac{1}{2})$ and:

$\ell' = 1 \; (\ell' = [1\text{-}2])$

for which j' = 2, 1 and 0. A 'triplet' spectral line appears – which might be depicted (with an appropriate diagram) with energy states $^3P_2$, $^3P_1$, and $^3P_0$. The capital letter (P) denotes one of several energy levels given by S, P, D and F, etc., corresponding to values of the quantum number ℓ = (0, 1, 2, 3 etc.) in an ascending alphabetical order from F.

Hence, ℓ' = 1 implies P-level, and the subscripts denote J' corresponding to 0, 1, 2 values. The superscript '3' is the *multiplicity*, derived from the relation: $2s' = 2\ell' + 1$.

It is instructive, to understand atomic shell structure, to study the table below to see how the principal quantum number, n, changes with ℓ and the subshell as well as the symbol for the primary electron shell. The Table shown below shows the respective Shell and Subshell Symbols and associated Quantum Numbers

n =	SHELL	ℓ =	Subshell
1	K	0	s
2	L	1	p
3	M	2	d
4	N	3	f
5	O	4	g
6	P	5	h

In working out assorted problems, preparation of a schematic energy level diagram associated with the state of the system can also be of immense value.

End of Chapter Problems:

1) The activity A of a radio-nuclide is given as:

$$A = A_0 \ exp \ (-\lambda t)$$

Where A_0 is the decay rate at time $t = 0$, and A refers to the decay rate at some time t thereafter. If a particular radio-nuclide has $A_0 = 1.1 \times 10^{10}$ decays/sec and a half life $T_{1/2} = 28.0$ years, find:

a) the decay constant, λ,

b) The activity A after 1 hour, after 2 hours.

c) The activity A after 49 years,

d) The number of radioactive nuclei after 49 years if the original number $N_0 = 2.5 \times 10^{17}$

2) A point source of gamma radiation has $(T\frac{1}{2}) = 30$ mins. The initial count rate recorded by a G-M tube is 360/s. Find the count rate that would be recorded after 4 half lives. Sketch the decay curve and determine the activity, A.

3) At a certain instant, a sample of a radioactive material contains 10^{12} atoms. The half-life of the material is 30 days.

a) Calculate the number of disintegrations in the first second.

b) The time elapsed before 10,000 atoms remain.

c) the count rate corresponding to the time in (b).

4) A radionuclide sample of N = 10¹⁵ atoms undergoes decay at the constant average rate of **dN/dt = 6.00 x 10¹¹ /s.** From this information, find:

a) The Activity A

b) The decay constant λ

c) The half life of the sample in minutes.

5) Using Fig. 5 as a basis, compute the energies and wavelengths of the photons emitted when the electron in the hydrogen atom makes the 1st, 2nd and 3rd Balmer transitions.

6)In a given experiment, the work function for a particular metal is known to be ϕ = 3.1 eV. In a single trial radiation of wavelength 270 nm is allowed to fall on the plate of the anode surface. Use this to find the stopping potential used in the experiment, and the threshold frequency.

7) Assume an experiment is done in which the light intensity is kept constant *but 3 different frequencies are used to elicit photo-electrons from the same metallic surface.* If each trial employs a different stopping potential with: V3 > V2 > V1, then sketch a graph of the results, with photo-current plotted vs. stopping potential

7)Draw to scale the possible orientations – in a strong magnetic field – of an orbital angular momentum vector with ℓ = 2.

8) Find the possible values of the total angular momentum for an electron with $\ell = 0, 1,$ and 2.

9) Find the maximum possible total angular momentum quantum number for a system with 2 p electrons.

10) For a hydrogen atom in the $\ell = 3$ state, calculate the magnitude of the orbital angular momentum, and the allowed values of L_z and θ.

11) Find the possible values of s', ℓ' and j' for an atomic configuration with 2 optically active electrons and quantum numbers: $\ell_1 = 2$, $s_1 = \frac{1}{2}$, $\ell_2 = 3$, $s_2 = \frac{1}{2}$.

12) For the *4s 3d* configuration we have:

$\ell_1 = 2$, $s_1 = 1/2$, $\ell_1 = 0$ and $s_1 = 1/2$.

Using the assorted combinations, for $\ell' = 0$ and $\ell' = 2$, to get the respective j' values (in combination with s' = 0), and then further for s' = 1, sketch the energy configuration diagram

13) In the case of **Krypton**, why is it that the [Kr] $4d^9$ $5s1$ state has a higher energy than the [Kr] $4d^9$ state?

14) A two-electron atom for which the orbital angular momentum quantum numbers are $\ell_1 = 3$ and $\ell_2 = 2$ can have what values for the total orbital angular momentum number L? Determine the possible values of the total angular momentum quantum number of single **f** electron.

Chapter III. Basics of Quantum Theory

3.1. The Wave Model of the Atom.

Though useful, especially in terms of identifying spectral lines, Bohr's model had its limitations. For example, it couldn't account for how angular momentum is conserved in atoms, nor how electronic transitions originate. More seriously, it wasn't able to deal with the problem of lost energy and why atoms don't simply collapse.

Consider the following dynamical picture: as the electrons whir about the nucleus they ought to be *losing energy*, in the context of Bohr's orbital model. If they lose kinetic energy over time they must spiral into the nucleus, and the atom then ceases to exist. This ought to happen in a very short time, so that most atoms in the universe cease to exist and hence the whole universe. But this isn't observed. Why?

The only explanation is that Bohr's orbital model can't be correct. Thus was born the theoretical basis for the **wave model** which we mostly accept today in modern quantum mechanics. Unlike the Bohr model, electrons don't follow *defined orbital paths* but instead are referenced to regions *or volumes* in which they will be more or less probable. The basic allocation of electrons, say for the hydrogen atom, is then confined to "orbitals" or regions of higher probability.

We now look at the experimental basis provided for this model.

Around 1926, a young French physicist named **Louis de Broglie** actually postulated the basis for material particles, such as electrons, acting *as waves*. This was experimentally verified in the (1927) Davisson and Germer electron diffraction experiment sketched below:

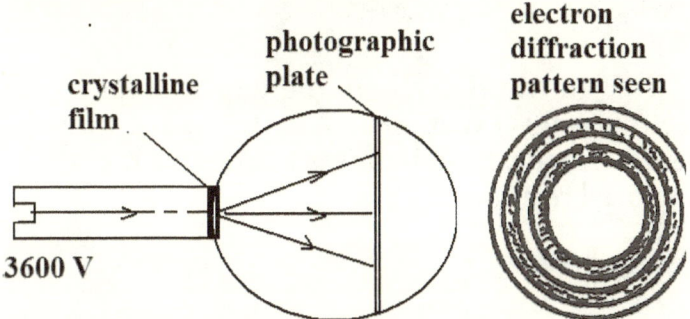

Fig. 3.1: The Davisson and Germer experiment

From the experiment, with electrons moving through a potential difference V = 4,000 volts, the kinetic energy gained should be equal to the work done, or:

$$\tfrac{1}{2} m v^2 = eV$$

Where m the mass of the electron is: 9.1 x 10 $^{-31}$ kg

And the electron charge e = 1.6 x 10^{-19} C

The velocity then is:

$$v = \sqrt{(2eV/m)}$$

The momentum p = mv = m $\sqrt{(2eV/m)}$ = $\sqrt{(2eVm)}$

And the de Broglie wavelength is:

$$\lambda_D = h/p = h/\sqrt{(2eVm)}$$

For a voltage V = 3,000 V one would find: λ_D =

$(6.626 \times 10^{-34}$ J-s)$/ [(3.2 \times 10^{-19}$ C$) (3 \times 10^3 V) (9.1 \times 10^{-31}$ kg$)]^{1/2}$ $= 2 \times 10^{-11}$ m

Which is the ***de Broglie wavelength*** of the electron in this experiment. A first step to uncovering the wave model from Bohr's is to examine his quantized relationship:

m vr $= nh/ 2\pi = L$

where L is the angular momentum. Re-arranging:

h/ mv $= 2\pi$ r$/$ n $= \lambda_D$ or 2π r $=$ n λ_D

Showing the radius is scaled into n (standing) waves of wavelength λ_D as shown below:

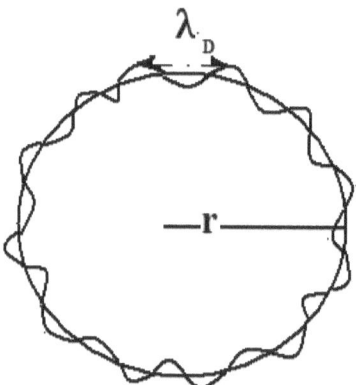

Fig. 3.2. Standing wave model for the Bohr atom.

This wave-orbiting electron atom still has a radius r, but with waves each separated by one de Broglie wavelength , λ_D. Thereby an integral number of such wavelengths form the circumference of the atomic orbit, as required by the condition: $2\pi r = n \lambda_D$.

For example, in the case of hydrogen the first three of these cloud-wave regions is shown in Fig. 3.

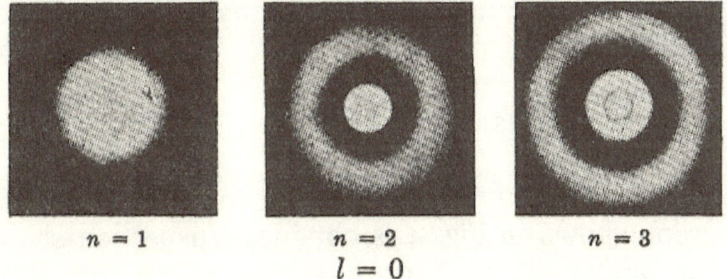

$n = 1$ $n = 2$ $n = 3$
$l = 0$

Fig. 3.3: Electron cloud-regions in Hydrogen

Let us 'zoom in' on the more spherical n= 1 configuration, and the probability for the election in this space as depicted in Fig. 4 below:

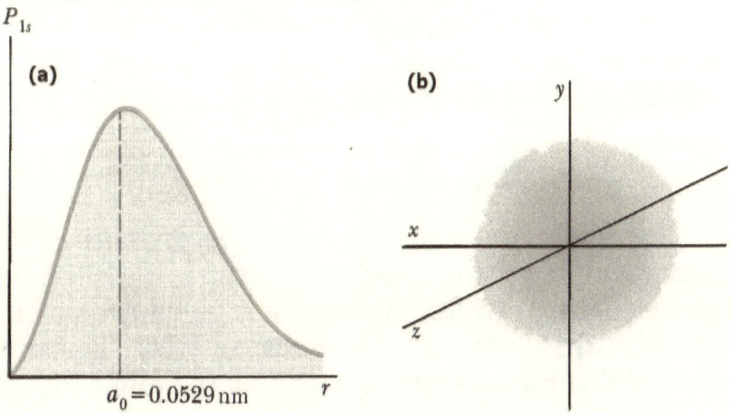

Fig. 3.4: The n = 1 electron orbital for hydrogen

This diagram, more than any other, dispenses with the notion that hydrogen electron occupies a definite position. Instead, it's confined someplace within a "cloud" or probability (b) but that probability can be computed as a function of the Bohr radius (a_o = 0.0529 nm). The probability P_{1s} for the 1s orbital is itself a result of squaring the "*wave function*" for the orbital. If the wave function is defined ψ (1s) = $1/\sqrt{\pi}$ (Z/a_o) exp ($-Zr/a_o$), and the probability function is expressed:

$$P = |\psi (1s) \psi (1s) *|$$

Where ψ (1s) * is the complex conjugate, then the graph shown in Fig. 4 is obtained. Inspection shows the probability of finding the electron at the Bohr radius is the greatest, but it can also be found at distances less than or greater than 0.0529 nm.

We thereby see from Figs. 3-4 that Bohr's original quantizing number, n, has far more meaning than simply to parse the number of standing waves for a given atom. We already see that it determines the energy of the atom, viz.

$$E_n = -13.6/\ n^2$$

But it also indicates *the average distance of the electron from the nucleus.* Thus, de Broglie's wavelength provides the basis for the wave-particle duality that lies at the basis of the "smeared" probabilistic atoms peculiar to modern atomic theory.

At the bottom of Fig. 3 are the quantum numbers: n and ℓ, which are identified as: the principal quantum number, and the angular momentum quantum number, respectively.

There are two physical meanings attendant on n: i) it determines the energy of an orbital (specifically in the H-atom), and (ii) it indicates the average distance of an electron in a particular orbital, to the nucleus, To fix ideas, I show in the accompanying diagram (Fig. 5) a sketch of one lobe for an electron orbital associated with the (3, 2, ±2) state in the Hydrogen atom. The key point is the orbital denotes an electron density associated with a probability of finding the electron in some defined space.

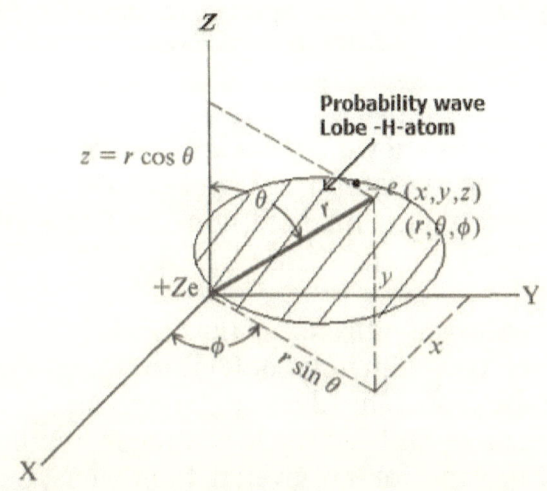

Fig. 3.5: one lobe for an electron orbital associated with the (3, 2, ±2) state in hydrogen

In the case shown one must also visualize a symmetrical lobe on the other side (making the whole orbital resemble a dumbbell) to make it complete.

As one alters the set of quantum numbers the electron densities change and so do the probabilities associated with the orbit. (See Fig.6)

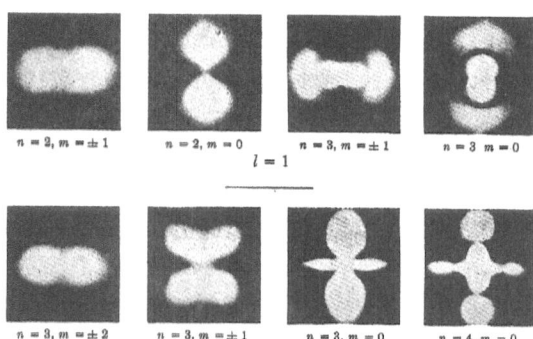

$n = 2, m = \pm 1$ $n = 2, m = 0$ $n = 3, m = \pm 1$ $n = 3 \;\; m = 0$

$l = 1$

$n = 3, m = \pm 2$ $n = 3, m = \pm 1$ $n = 3, m = 0$ $n = 4, m = 0$

Fig. 3.6: Further hydrogen orbitals with higher quantum numbers

Describing orbitals using the set of given quantum numbers means knowing the numbering rules applied to each. In the case of the principal quantum number, n, we allow it to have integral (non-zero) values: 1, 2, 3, 4 etc.

The physical significance of the angular momentum quantum number (ℓ) is to convey the shape of the probability density cloud or orbital. The numbering rule for l is directly contingent on the value for n. Thus, for any given n, then ℓ must be such that it has integral values from 0 to (n -1). This means if n = 2, then ℓ can have (n- 1) = (2 -1) = 1. But if n =1, then ℓ = (n - 1) = 1 - 1 = 0.

Note that the ℓ -quantum numbers appear more than once for any orbital with n >1. Thus, for the n = 2 case, we have two values of ℓ occurring: one for 1 = 0, the other for ℓ = 1. If we go on to n= 3 there are three values of ℓ, for n = 4, four values and so on. One also finds the ℓ -value specified for lettered orbitals: *s, p, d, f, g, h*. The s-orbital is for ℓ =0, the p for ℓ =1, the d for ℓ = 2 and so on.

There is no special significance to the letters (apart from the physical meaning we already gave for the angular momentum quantum number, ℓ, and they are mainly of historical import- though still retained, for example, in chemistry. (By extension, one also often hears the term "atomic shell" used in chemistry). A collection of orbitals under the same value of n is called a "*shell*". Thus, for n = 4, we have ℓ =0, ℓ = 1, ℓ =2, ℓ = 3 so comprising the collection of orbitals: s, p, d and f.

Lastly, there is the *magnetic quantum number*, usually designated m_ℓ (subscript the same as the angular momentum quantum number) because it is contingent upon it. This quantum number describes the orientation of the orbital in 3-D space. For a given angular momentum quantum number, ℓ, we have integral values specified as follows:

$m_\ell = -\ell, (-\ell +1)....0......(\ell - 1), +\ell$

Note the above set of m_ℓ numbers is given as a *SERIES*, e.g. starting with $(-\ell)$ and terminating at $+\ell$. Look at the simplest example for ℓ = 0, then:

$m_\ell = 0$.   (Since all terms are zero)

What about ℓ = 1?

Then: $m_\ell = -1, 0, 1$

What about $\ell = 2$?   We have:

$m_\ell = -2, -1, 0, 1, 2$

As a general rule then, we can use the formula:

$N(m_\ell) = \{(2 \times l) + 1\}$

to give the total number of m_ℓ numbers.

3.2. Non-Commuting Observables and the Heisenberg Uncertainty Principle

The principle of non-commuting observables and their relation to the Heisenberg Uncertainty principle is often ignored in many modern physics approaches and courses. What are non-commuting observables?

In modern quantum mechanics each probability-bearing proposition of the form *"the value of physical quantity A lies in the range B"* is represented by a projection operator on a Hilbert space $\mathbf{H}$[1].   John von Neumann [1932], showed that each physical system can be associated with a (separable) *Hilbert space* $\mathbf{H}$, the unit vectors of which correspond to *possible physical states of the system.*

Each "observable" real-valued random quantity can then be thought of as a property of the system which may be measured or not. It is usually represented by

---

[1] This is defined as a linear vector space V with an infinite number of dimensions, i.e. $V = v1 + v2 +$ ....vn where n = ∞. In quantum mechanics it is assumed Hilbert spaces are used unless otherwise noted.

what is called a self-adjoint operator $\hat{A}$ on **H**, the spectrum of which is the set of *possible values* **of** $\hat{A}$. If **e** is a unit vector in the domain of A, representing a *state of the system*, then the expected value of the observable represented by $\hat{A}$ in this state is given by the inner product . The observables represented by two operators **A** and **B** are commensurable if and only if **A** and **B** commute, i.e., **AB = BA**.

Thus, non commutation is possible, and indeed even expected, hence the emergence of the **Heisenberg Uncertainty Principle** which is really a statement regarding non-commutativity. This can be expressed in quantum mechanics, using the momentum (p) and position (x) measurements via the Poisson brackets:

$[x, p] = -i\hbar$

Where $\hbar$ is the Planck constant of action h, divided by 2π.

If two variables a, b commute, then one has:

$[A, B] = (A \bullet B - B \bullet A) = 0$

If not, then:

$[A,B] = (A \bullet B - B \bullet A) = -1$

and we say a and b are *'non-commuting'*.

(I.e. you may observe one aspect but not the other).

In term's of Bohr's **Complementarity Principle**, the variables x (position) and p(momentum) are regarded as "*mutually interfering observables*".

This is why only one observable can be obtained to precision, while you lose the other. Simultaneous precision in measuring two observables isn't allowed and this is the basis of the Heisenberg Uncertainty principle – which effectively imposes limits on the degree of measurement accuracy.

The form for the Heisenberg Uncertainty Principle, say in one dimension, is:

$$\Delta x \, \Delta p_x \approx h$$

Which is analogous to: $[x, p] = -i\hbar$

Where Δx is the uncertainty in position in the x-direction and Δp_x is the uncertainty in momentum.

One can't obtain simultaneously accurate values of both p and x since if he estimates or knows one to perfection, say Δp_x, he loses all information on the other (Δx) Thus, rewriting the Heisenberg relation (in 1-dimension) to find the uncertainty in momentum:

$$\Delta p_x \geq h / \Delta x$$

And, if Δp_x is known to perfect accuracy we have $\Delta x = 0$, then:

$$\Delta p_x \geq h / 0 = \infty$$

Hence no accurate measure is possible incorporating Δx because the uncertainty in their position is now infinite. The Heisenberg Energy-Time Uncertainty Principle is also used, and is expressed by:

$$\Delta E \, \Delta t \geq h/2\pi$$

3.3. *The Wave-Particle Duality*

We now look in somewhat more detail at wave-particle duality as it arises in quantum mechanics. In the particle interpretation, electrons fired from a device such as an electron gun would not all follow the same path since the trajectory of an electron – unlike a missile- can't be predicted from its initial state.

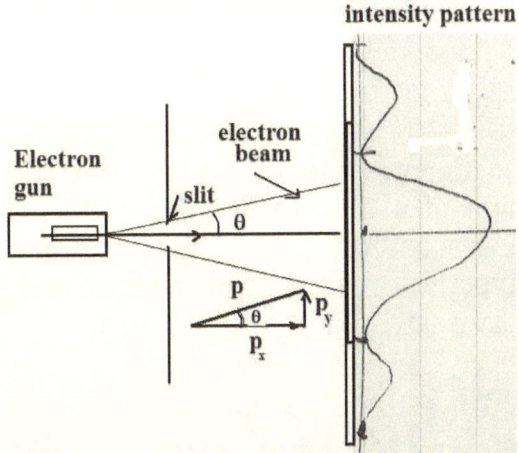

Fig. 3.7: *Showing electron diffraction, intensities*

We consider here the case of electron diffraction, based on Fig. 7, in which electrons are emitted from an electron gun and pass through a slit toward a detector or photographic plate onto which a

87

diffraction pattern appears. This pattern will also coincide with an intensity distribution such as shown. In effect, the intensity distribution basically describes the probability for an individual particle (electron) to strike each of several areas designated on the photographic film. This discloses a fundamental indeterminacy that has no counterpart in Newtonian mechanics. Now, consider an electron striking at some angle θ, such as indicated:

We have, from the quantities shown:

$p_y / p_x = \tan \theta$ or $p_y = p_x \theta$ (in limit of small θ)

Therefore, the y-component of momentum can be as large as:

$p_y = p_x (\lambda / a)$

Where a denotes the slit width. The narrower the dimension of a the broader the diffraction pattern, and the greater Δp. From Louis de Broglie's matter wave hypothesis (already introduced into the Bohr atom, as we saw, cf. Fig. 7) : $\lambda_D = h / p_x$

Therefore:

$p_y = p_x (h / p_x a) = h / a$ or: $p_y a = h$

But 'a' represents uncertainty in electron position vertically (Δy), i.e. as it passes through the slit. We can reduce Δp_y only by narrowing the slit width a and vice versa. Thus we get:

$$p_y a = \Delta p_y \, \Delta y \approx h$$

Which is one form of *the Heisenberg Uncertainty Principle* which states that the momentum of a quantum particle and its positions cannot simultaneously be known to the same arbitrary precision. One corollary is that to detect a particle any given detector must interact with it thereby altering the motion of the particle.

This view is no longer taken in any literal way because we understand that the quantum measurements are statistical in nature and hence a particular measurement is the result of a vast statistical assembly. Paul Dirac, in his book. defined the "principle of superposition" thusly[2]:

"A state of a system may be defined as a state of undisturbed motion that is restricted by as many conditions or data as are theoretically possible without mutual interference or contradiction"

Let's examine this in more detail. By "undisturbed motion" Dirac meant the state is pure and hence no *extraneous observations* are being made such that the state experiences interference effects to displace or disturb it. In the Copenhagen Interpretation, "disturbance" of mutually defined variables, say x, p or position and momentum, occurs only if:

$$[x, p] = -i\hbar = -i\,h/2\pi$$

---

[2] Dirac, P.A.M.: 1941, *Quantum Mechanics,* Oxford University Press, 11.

If it were the case that [x, p] = 0, one would say the variables "commute" and hence *there's no interference*. If the condition doesn't hold, then interference exists. Hence, Dirac's setting of an upper limit in the last portion of his definition, specifying as many conditions as theoretically possible *"without mutually interfering interference."* This state is undisturbed. We have a statistical perspective!

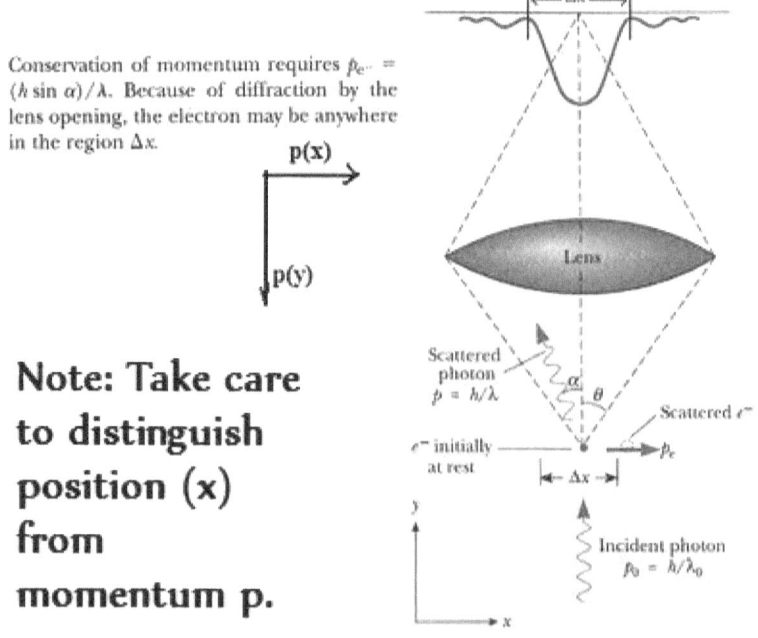

Conservation of momentum requires $p_e = (h \sin \alpha)/\lambda$. Because of diffraction by the lens opening, the electron may be anywhere in the region Δx.

Note: Take care to distinguish position (x) from momentum p.

Fig. 3.8: Sketch of Heisenberg Microscope and key parameters.

It is important to see from the preceding how the Heisenberg Uncertainty Principle arises not just from an *ad hoc* assumption, but from the limits (or "tolerance thresholds") of explicit quantities (e.g. p, x), when considered in the quantum limit. Hence, the model of the Heisenberg "microscope" provides a useful (although not practical, since it can't actually be constructed) means of deriving the statistical principle of superposition based on an observational ansatz.

Consider a measurement made to determine the *instantaneous position of an electron* by means of a microscope. In such a measurement the electron must be illuminated, because it is actually the light quanta (photon) scattered by the electron that the observer sees. The resolving power of the microscope determines the ultimate accuracy with which the electron can be located. This resolving power is known to be approximately:

$\lambda / 2 \sin \theta$

Where λ is the wavelength of the scattered light and θ is the half-angle subtended by the objective lens of the microscope. Then:

$\Delta x = \lambda / 2 \sin \theta$

In order to be collected by the lens, the photon must be scattered through any range of angle from $-\theta$ to θ. In effect, the electron's momentum values range from:

$h \sin \theta / \lambda$ to $- h \sin \theta / \lambda$

Then the uncertainty in the momentum is given by:

$\Delta p_x = 2 h \sin \theta / \lambda$

Then the Heisenberg Uncertainty Principle product is:

$\Delta p_x \ \Delta x \ = \ (2\,h \sin \theta/\lambda \ ) \ (\lambda/\,2\sin\theta) \ = h$

3.4. *Probability density and Expectation Values*

Earlier we saw:

$P = \left| \psi\,(1s)\ \psi\,(1s)\ ^* \right|$

Which is the probability density and a quantity we can actually measure, e.g. for the 1s state of hydrogen. Then this needs to be generalized to apply to more than one case.

Since the electron locations can't be computed from Newtonian mechanics but more plausibly based on an analogous probability density to what we saw above, then we can generalize and write:

$P_{ab} \ = \ \int^b_a \ \| \psi(x) \|^2 \ dx$

Where x is the state under consideration and this system is 1-dimensional with the probability assessed from a to b.

Note that we define the *normalization condition* as:

$\int^b_a \ \|\psi\|^2 \ dx \ = 1$

Normalization is simply a condition stating that the particle exists at some point at all times. Thus if we had:

$$\int {}^b_a \ \|\psi\|^2 \ dx \ = \ 0$$

The probability would not exist. The probability condition then allows us to specify the probability of observing a particle even though we cannot specify the position. The normalization then gives the probability of finding the particle in the range a $\leq$ x $\leq$ b, say in one dimension.

The Schrodinger wave function, $\psi(x)$ satisfies the *Schrodinger equation*. In one dimensional, this differential equation is:

$$d^2 \ \psi/dx^2 + F(x) \ \psi = 0$$

Though the function $\psi(x)$ is not a measurable quantity, other measurable quantities such as the energy E and momentum of the particle can be derived from it. Also, if the wave function is known it is possible to compute the average position of the particle, known as the *expectation value*:

$$<x> \ = \ \int {}^\infty_{-\infty} \ x \ \|\psi(x)\|^2 \ dx$$

This expression implies the particle is in a definite state so that the probability density is time – independent.

Worked Problem:

Consider the 1D 'box' shown below, and a particle confined therein:

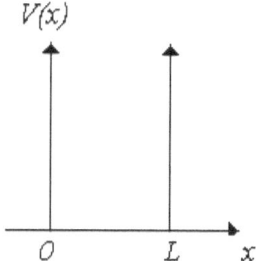

$V(x)$

0 L x

With maximum dimension L in direction +x. Find the probability P $_{ab}$ the particle is between a = o and b = L, the expectation value <x> and also show the energy for the particle can be quantized according to:

$E_n = (h^2/ 8m L^2) n^2$

Let the wave function be: $\psi(x) = \sqrt{2}/ \sqrt{L}$ $\sin (kx)$

Solution:

Rewrite the wave function as:

$\psi(x) = \sqrt{2}/ \sqrt{L}$ $[\sin (\pi x/L)]$ where $k = \pi/L$

Then:

$P_{ab} =$

$\int_a^b |\psi(x)|^2 \, dx = \int_o^L (\sqrt{2}/ \sqrt{L})^2 \, \sin^2 (\pi x/L) \, dx$

$= 2/ L \int_o^L \frac{1}{2} [1 - \cos (\pi x/L)] \, dx$

(Let $\theta = \pi x/L$ and use: $\sin^2 \theta = \frac{1}{2} (1 - \cos 2\theta)$)

$$P_{ab} = 2/L[\tfrac{1}{2}\int L_o \, dx - \int L_o \cos(\pi x/L)]\,dx$$

$$P_{ab} = 1 - 1/\pi \sin(2\pi x/L)]^{L_o} = 1 - 1/\pi \sin(2\pi)$$

But : $\sin(2\pi) = 0$ so $P_{ab} = 1$

The expectation value is:

$$<x> = \int_{-\infty}^{\infty} x\,|\psi(x)|^2 \, dx$$

$$<x> = 2/L[\int L_o \, x\sin(\pi x/L)^2]\,dx$$

$$<x> = L^2/4 - [x\sin(2\pi x/L)/4\pi/L - \cos(2\pi x/L)/8(\pi^2/L^2)]$$

$$<x> = 2/L\,(L^2/4) = L/2$$

To find the energy we have the Schrodinger equation:

$$d\psi^2/dx^2 + K^2\psi = 0$$

where $K = \sqrt{[2mE]}/\hbar$

If we examine the sketch below:

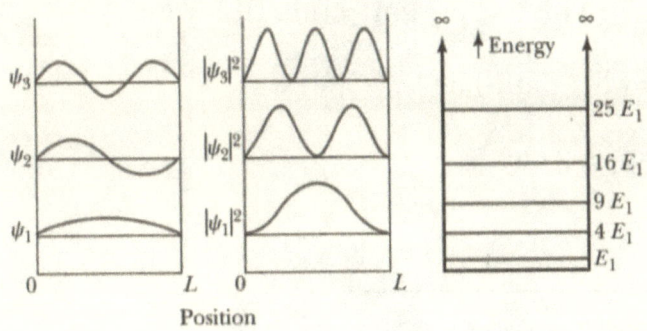

Position

We see plots of the wave function $\psi(x)$ vs. position x (far left), and of the probability density (middle) and the energy levels. Since we have represented the wave function by a sinusoidal function then it follows that the allowed wavelengths are those for which the length L is equal to an integral number of half wavelengths, or:

$$L = n\lambda/2$$

These allowed states are called stationary states and represent standing waves (analogous to the ones seen earlier for the Bohr atom). Thus, the wavelengths of the particle are restricted by the condition:

$$\lambda = 2L/n$$

Then the magnitude of the momentum p is also restricted to specific values (e.g. using $p = h/\lambda$) [3] such that:

$$p = h/\lambda = h/2L/n = nh/2L$$

The energy associated with the particle is then:

$$E = 1/2\, mv^2 = p^2/2m = (nh/2L)^2/2m$$

$$E = (h^2/8mL^2)\, n^2$$

n= 1, 2, 3 etc.

---

[3] Recall from Planck's law: $E = hc/\lambda$ and $p = \sqrt{[2mE]}$.

Thus the energy is quantized with the energy of the lowest energy state corresponding to n =1 so:

$$E_1 = (h^2/ 8mL^2)$$

This least energy that the particle can have is called the "zero point energy" and means the particle can never be at rest.

Note that the above energy result can also obtained through the use of differential equations, for which one solution will be provided in the Appendix (Appendix IV).

The probability density can be extended to 3 dimensions by writing, with dV the volume:

$$P = \int_{-\infty}^{\infty} |\psi(x)|^2 \, dV$$

The quantized energy will be (for a 3D box):

$$E = (h^2/ 8mL^2)[n_x^2 + n_y^2 + n_z^2]$$

End of Chapter Problems:

1)For a 1D box, let an electron inside have a wave function:

$$\psi(x) = \sqrt{2}/ \sqrt{L} \; [\sin (2\pi x/L)]$$

Find the probability of locating the electron between x = 0 and x = L/4. (Can be left in integral form).

2)Use the uncertainty principle to estimate the uncertainty in momentum for a particle in a 1D box. Estimate the ground state energy using this means and compare it to the actual ground state energy.

3) The wave function for a particle confined to moving in a 1D box is given by:

$$\psi(x) \; = \; A\,[\sin{(n\pi x/L)}]$$

Use the normalization condition on $\psi(x)$ to show the constant A is given by: $A = \sqrt{2}/\sqrt{L}$

4) It is known from quantum mechanics that a particle in a one dimensional potential well (such as shown in the diagram) can exist in a number of energy states. Imagine an electron confined between the boundaries x and $x + \Delta x$, where Δx is 0.5 Angstroms.

Approximately, what is the uncertainty in the x-component of the momentum of the electron?

5)(a) Consider a free particle confined between two impenetrable walls at x and $x + L$. What is the probability according to classical physics that the particle will be found between x and $x + L/3$ *if no other information is given*?

b) What is the probability according to quantum mechanics that the particle in its *lowest energy state* will be found between x and $x + L/3$?

c) What is the probability according to quantum mechanics that the particle in the **second lowest energy state** will be found between x and $x + L/3$

Chapter IV. Operators and Complex Quantum Systems

4.1. Writing the Hamiltonian:

The brilliance of the early quantum mechanicians lay in substituting the operators (E_{op}, p_{op}) for the corresponding quantities of the original classical Hamiltonian, then multiplying through by the wave function ψ :

$$H_{op} \, \psi \; = \; E_{op} \, \psi$$

In this way, a drastically simplified basic quantum mechanical equation could be obtained, which could then be expanded once one substituted the operators, i.e.:

$$p_{op} \; = \; -i \, \hbar \, (\partial / \partial x)$$

$$E_{op} \; = \; i \, \hbar \, (\partial / \partial t)$$

So, the full wave equation becomes:

$$\boxed{- \hbar / 2m \, (\partial^2 / \partial x^2) \, \psi \; + \; V(x) \, \psi \; = \; i \, \hbar \, (\partial / \partial t) \, \psi}$$

A more common form of the above equation for less advanced work, say as applied in Calculus Physics courses is the 1-dimensional form of the Schrodinger equation, which is not time dependent but time-**independent**.:

$$d^2 \psi / dx^2 + 8\pi^2 \, m_e / h^2 \, \{W - V(x)\} \psi = 0$$

Where W is *the total energy* of each electron so (W - V) is the kinetic energy, i.e.

$W = V + [m_e v^2/2] = V + KE$

so: $KE = W - V = [m_e v^2/2]$, thence:

$2 m_e (W - V) = [m_e v^2/2]$,

Then:

$m_e{}^2 v^2 / h^2 = 2 m_e (W - V)/ h^2$

To illustrate some basic properties of the 1-dimensional Schrodinger equation consider the step function.. Consider then a particle moving freely as shown with energy E toward a quantum energy 'step':

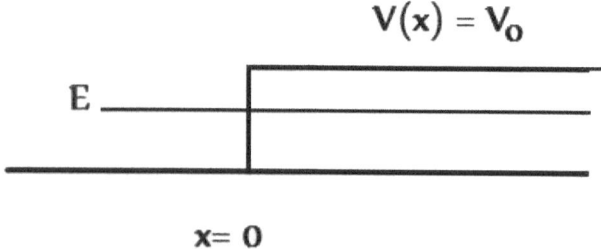

Fig. 4.1: Diagram for the simple step function

Such a "free particle" is not bound to any potential, so in this specific case, we have that $V(x) = 0$, so:

$d^2\psi/dx^2 + 8\pi^2 m_e /h^2 \{W\} \psi = 0$

Another interesting facet of the Schrodinger equation refers to the **superposition** aspect. If we start, say, with two different initial conditions, to obtain two waves:

$$\psi = \psi_1(x)$$

and $\psi = \psi_2(x)$

Then *all solutions* of the given Schrodinger wave equation are of the form:

$$\psi = A\,\psi_1(x) + B\psi_2(x)$$

In the task of understanding quantum mechanics it is useful to see the working from actual examples. As I've learned in teaching the subject, students are apt to get much more out of it if several different examples of quantum systems are explored and solved.

To that end, we now consider a simple QM system set up such as shown in the sketch below where we have a beam of electrons of energy kinetic energy E incident on a plane where there is a potential step such that E is less than V(x).

Since the total energy E is a constant, then by classical Newtonian mechanics the electron cannot enter the region at $x > 0$. We must have then:

$$E = p^2/\,2m + V(x) < V(x) \ \text{or:} \ p^2/\,2m \ < 0$$

To determine the motion of the electron quantum mechanically we must find the wave function that is a solution for the total energy $E < V_o$. In one region the appropriate time – independent Schrodinger equation is:

$-h^2/2m\ [d^2\psi(x)/dx^2] = E\ \psi(x), \quad x < 0$

In the other region, for which $x > 0$ we need:

$-\quad h^2/2m\ [d^2\psi(x)/dx^2] + V_o\ \psi(x) = E\ \psi(x)$

These two equations can be solved separately. An eigenfunction for the entire range of x is then obtained by joining the two solutions together at $x = 0$ in such a way to satisfy the key conditions, i.e. that $d\psi(x)/dx$ must be finite and continuous:

The first equation is simply that for a free particle, so the general solution can be written in the traveling wave form:

$\psi(x) = A\ \exp(iK_1\ x) + B\ \exp(-iK_1\ x)$

Where: $K_1 = \sqrt{(2mE)}\ /\ \hbar$

For the 2nd equation the general solution would be:
$\psi(x) = C\ \exp(iK_2\ x) + D\ \exp(-iK_2\ x)$

Where: $K_1 = \sqrt{(2m(V_o - E))}\ /\ \hbar$ **for x > 0**

Continuity of $\psi(x)$ is satisfied if the relation:

$D\ \exp(-iK_2\ x)_{x=0} =$

$A\ \exp(iK_1\ x)_{x=0} + B\ \exp(-iK_1\ x)_{x=0}$

Is satisfied. If it is, then:

$D = A + B$

Continuity of the derivatives of the solutions is expected if:

$- K_2\, D \exp(-iK_2\, x)_{x=0} =$

$i\, K_1\, A \exp(iK_1\, x)_{x=0} - iK_1\, B \exp(-iK_1\, x)_{x=0}$

In the latter case: $iK_2D/\, K_1 = A - B$

Adding the two equations in D, A and B:

$A = D/\, 2\, (1 + iK_2/K_1)$

Subtracting gives:

$B = D/2\, (1 - iK_2/K_1)$

Then the eigenfunction for this potential will be:

For $x \leq 0$:

$\psi(x) =$

$D/\, 2(1 + iK_2/K_1)\, (e^{\,iK_1x}) + D/2\, (1 - iK_2/K_1)\, e^{\,iK_2x}$

$\psi(x) = D \exp(-iK_2\, x)$ for $x \geq 0$

The preceding result amounts to a transmitted and reflected mode and can be interpreted using the probability flux, $S(x,t)$ in the region $x < 0$. According to the de Broglie wave postulate:

$p_1 = mv = \hbar K_1$

And the probability flux is: $S(x,t) = v\,A^*A = v\,P(x,t)$

But: $P(x,t) = B^*B$ so that we can thereby obtain the expression:

$S(x,t) = v\,A^*A - v\,B^*B$

Where $v = \hbar K_1/ m$

The first term in the $S(x,t)$ expression originates from the probability flux flowing in the direction of increasing x. The second or 'B-term' originates from the probability flux flowing in the opposite direction. As a result we can associate the first term (in the region $x < 0$) with the incidence of the particle on the point where the potential energy $V(x)$ changes and the second term with the reflection of the particle from the change in potential. The computation of the intensity of the reflected probability flux to the incident intensity can then be made:

$R = v\,B^*B/ vA^*A = B^*B/ A^*A =$

$[(1 - iK_2/K_1)^* (1 - iK_2/K_1)] / [(1 + iK_2/K_1)^* (1 + iK_2/K_1)]$

$=$
$[(1 + iK_2/K_1)^* (1- iK_2/K_1)]/ [(1 + iK_2/K_1)^* (1 - iK_2/K_1)]$
$= 1$

The transmitted probability flux can be found by calculating $S(x,t)$ at some point in the region $x > 0$:

$S(x,t) = v_2\,C^*C$

Where $v2 = \hbar\, K2/\, m = p2/m$

The ratio of the intensity of the reflected flux to the incident flux is the probability the particle will be reflected back to the region x < 0. This is written:

$R = v1\, B^*B/\, v1\, A^*A = (K1 - K2)^2/\, (K1 + K2)^2$

The ratio of the intensity of the intensity of the transmitted flux to the intensity of the incident flux is the probability that the particle will be *transmitted* into the region x > 0 or:

$T = v2\, C^*C/\, v1\, A^*A = K2\, (2\, K1)^2/\, K1\, ((K1 + K2)^2$

$= 4K1\, K2/\, (K1 + K2)^2$

And it is easy to show then that: R + T = 1

4.2.Particle in a 3-Dimensional Square Well

One of the more interesting applications is for a particle confined to a 3-dimensional square 'well' or box, as shown:

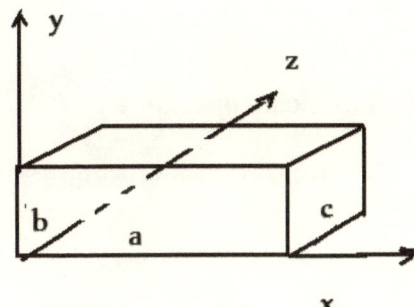

Fig.4. 2: Sketch for 3-dimensional box

Using this we can set out the rectangular regions defined according to:

$0 < x < a$

$0 < y < b$

$0 < z < c$

The potential energy inside the region is zero and outside is infinite so the Schrodinger equation becomes:

$$\mathbf{T}_{op}\, \psi \;=\; \mathbf{E}_{op}\, \psi$$

Where $\mathbf{T}_{op}$ is the kinetic energy operator which is written in 3D as:

$$\mathbf{T}_{op} \;=\; \mathbf{p}^\wedge\, \mathbf{p}^\wedge /\, 2m \;=$$

$$[\mathbf{p}^\wedge{}_x{}^2 + \mathbf{p}^\wedge{}_y{}^2 + \mathbf{p}^\wedge{}_z{}^2\,]\,/\, 2m$$

$$= -\hbar^2/\, 2m\,[\,\partial^2/\partial x^2 + \partial^2/\partial y^2 + \partial^2/\partial z^2\,) =$$

$$-\hbar^2/\,2m\, \nabla^2$$

Where ∇ is the Laplacian operator.

The Schrodinger equation then becomes:

$$\partial^2\psi/\partial x^2 + \partial^2\psi/\partial y^2 + \partial^2\psi/\partial z^2 + 2mE/\,\hbar^2\,\psi = 0$$

Solution of the 3D Square Well Schrodinger Equation:

This is a partial differential equation easily solved by the separation of variables:

$$\psi = X(x)\, Y(y)\, Z(z)$$

Then:

$$x' = \partial X/\partial x \quad y' = \partial Y/\partial y \text{ and } z' = \partial Z/\partial z$$

This leads to the equation:

$$X''YZ + XY''Z + XYZ'' + 2mE/\hbar^2\, XYZ = 0$$

The required normalization equation is then:

$$\int_{a_0} \int_{b_0} \int_{c_0} \|\psi\|^2\, dz\, dy\, dx = 1$$

Dividing the Schrodinger equation by XYZ:

$$X''/X + Y''/Y + Z''/Z + 2mE/\hbar^2 = 0$$

We let:

$$X''/X = -\alpha^2,\ Y''/Y = -\beta^2,\ Z''/Z = -\gamma^2$$

With α, β and γ constants.

Then, we have:

$$x^2 + \alpha^2 x = 0,\ y^2 + \beta^2 y = 0,\ z^2 + \gamma^2 z$$

So the independent solutions will be:

$$x = \sqrt{(2/a)} \; \sin(n_x \pi x/a)]$$

$$y = \sqrt{(2/b)} \; \sin(n_y \pi x/b)]$$

$$z = \sqrt{(2/c)} \; \sin(n_z \pi x/c)]$$

where: $n_x = 1, 2, 3, 3$ etc.

Then we can write for the wave function:

$$\psi = (8/abc)^{1/2} \sin(n_x \pi x/a) \sin(n_y \pi x/b) \sin(n_z \pi x/c)$$

And further, to obtain the quantized energy, E:

$$n_x^2 \pi^2/a^2 - n_y^2 \pi^2/b^2 - n_z^2 \pi^2/c^2 + 2mE/\hbar^2 = 0$$

And hence:

$$E = \pi^2 \hbar^2/2m \, [n_x^2/a^2 + n_y^2/b^2 + n_z^2/c^2)$$

If the box is a cube, i.e. $a = b = c$:

$$E = \pi^2 \hbar^2/2m \, a^2 (n_x^2 + n_y^2 + n_z^2)$$

(N.B. for states higher in energy than the ground state, many levels-distinct states have the same energy.)

4.3. The Hydrogen Atom

The Hamiltonian for the hydrogen atom Schrodinger equation can also be written in concise form as:

$H_{op} \psi = E_{op} \psi$

For which we only need to know the operators in order to expand it to its proper representation. In this case:

$H_{op} = p\hat{}r^2/ 2m + \ell^{\wedge 2}/ 2mr^2 + V(r)$

And: $p\hat{}r = - i \hbar (1/r \, \partial^2 / \partial r^2)$

The operators are radically different because we are dealing with an atom and treating it along the lines of radial symmetry – so the operator must be a function of r. We also note the form for the spherical wave function: $\psi = \psi (r, \theta, \varphi)$

Again, this is to preserve spherical symmetry. The full Schrodinger equation for the hydrogen atom can then be written as:

$[pr^2/ 2m + \ell^2/ 2mr^2 + V(r)] \psi (r, \theta, \phi) = E\psi (r, \theta, \varphi)$

This equation can now be used, after further expansion of the assorted operators, to obtain the properties of the hydrogen atom, including its energy levels.

End of Chapter Problems:

1)Use the Heisenberg uncertainty principle to estimate the uncertainty in momentum for a particle in a 1D box. Estimate the ground state energy using this means and compare it to the actual ground state energy.

2) It is known from quantum mechanics that a particle in a one dimensional potential well can exist in a number of energy states. Imagine an electron confined between the boundaries x and x + Δx, where Δx is 0.5 Angstroms.

Approximately, what is the *uncertainty in the x-component* of the momentum of the electron?

3) Consider a free particle confined between two impenetrable walls at x and x + L. What is the probability according to classical physics that the particle will be found between x and x + L/3 *if no other information is given*?

b) What is the probability according to quantum mechanics that the particle in its *lowest energy state* will be found between x and x + L/3?

c) What is the probability according to quantum mechanics that the particle in the **second lowest energy state** will be found between x and x + L/3?

4) Consider a step potential, with boundary conditions:

x > 0   and E < V₀   with $K_1 = \sqrt{(2m(V_0 - E)}\,/\,\hbar$

a) Show the general solution of the appropriate Schrodinger equation would be:

$\psi(x) = C \exp(K_2 x) + D \exp(-K_2 x)$

b) For the transmitted (T) and reflected (R) flux of the step potential, show that:

R + T = 1.

5)Find the expectation value for the momentum of a particle with a wave function:

$U(x) = A \exp [i(\alpha - \alpha^2 \hbar\, t/\, 2m]$

6)Using the operators, $\mathbf{p\char94 r}$ and $\ell\char94 2$

$=$

$-\hbar^2/\sin^2 \theta[\sin \theta\, \partial/\partial \theta (\sin \theta\, \partial/\partial \theta) + \partial^2/\partial \varphi^2]$

Write out the full form of the Schrodinger equation for the hydrogen atom in spherical coordinates. Thence, obtain the final form such that $\mathbf{H}$ op $=$

$- [E - V]\, \psi\, (\mathbf{r, \theta, \varphi})$

Hint: Replace m by the reduced mass,

$\mu = mM/\, m+M$

6) In the form shown, the variables $\mathbf{r, \theta, \varphi}$ may be separated by letting:

$\psi\, (\mathbf{r, \theta, \varphi}) = R(r)\, \Theta\, (\theta)\, \Phi(\varphi)$

And performing suitable manipulations. Do this in a similar way to the approach used to solve for the 3D box (previous section) and then show the solutions which result are:

i) $\Phi(\varphi) = \exp (i\, m_\ell\, \varphi)$

ii) $\Theta(\theta) = \Theta_{\ell m} = \sin^{m\ell}\theta \, F_\ell |m_\ell| (\cos\theta)$

iii) $R_{n\ell} = \exp(-Zr/na_o)\,[Zr/a_o]^\ell \, G_\ell \,(Zr/a_o)$

Explain how your solutions help to account for the Table entries below:

Quantum Numbers

n	l	m_l	Eigenfunctions
1	0	0	$\psi_{100} = \dfrac{1}{\sqrt{\pi}}\left(\dfrac{Z}{a_0}\right)^{3/2} e^{-Zr/a_0}$
2	0	0	$\psi_{200} = \dfrac{1}{4\sqrt{2\pi}}\left(\dfrac{Z}{a_0}\right)^{3/2}\left(2 - \dfrac{Zr}{a_0}\right)e^{-Zr/2a_0}$
2	1	0	$\psi_{210} = \dfrac{1}{4\sqrt{2\pi}}\left(\dfrac{Z}{a_0}\right)^{3/2}\dfrac{Zr}{a_0}\,e^{-Zr/2a_0}\cos\theta$

7) In the case of the first solution (i) we demand that the function be single-valued and continuous so that:

$\Phi(\varphi + 2\pi) = \Phi(\varphi)$

So that:

$\exp[\,i\,m_\ell\,(\varphi + 2\pi)] = \exp(i\,m_\ell\varphi)$

where m_ℓ is the magnetic quantum number. Dividing by $\exp(i\,m_\ell\varphi)$ we obtain:

$\exp[\,i\,m_\ell\,(2\pi)] = 1$

Indicate the condition on m_ℓ for which this is satisfied.

8) *The solutions* $\Theta_{\ell m}$ are associated Legendre polynomials (see Mathematics Supplement) and require two quantum numbers, the magnetic and the azimuthal, **ℓ.** If ℓ is quantized and can only be a positive integer, show its acceptable values based on the values of m.

9) The solutions $R_{n\ell}$ are Laguerre functions which also require two quantum numbers, ℓ and n to identify acceptable solutions. (Note: n is the principal quantum number)

The eigenfunctions $\psi\,(\mathbf{r},\,\theta,\,\varphi)$ are formed by taking the product of the 3 types of functions so that:

$$\psi_{n\ell m\ell}\,(\mathbf{r},\,\theta,\,\varphi) = R_{n\ell}\,(r)\,\Theta_{\ell m}\,(\theta)\Phi_{m\ell}\,(\varphi)$$

By inspection of the table (Prob. 6) , identify the part or whole of the expression for the wave function ψ_{210} that comes closest to the form for $R_{n\ell}\,(r)$.

10) Complete the normalization condition for the H-atom:

$$\int_\infty{}_0 \int^\pi{}_0 \int^{2\pi}{}_0 \; \psi^*_{n\ell m\ell}\,(\mathbf{r},\,\theta,\,\varphi)\,\psi_{n\ell m\ell}\,(\mathbf{r},\,\theta,\,\varphi) \; dV = \underline{?}$$

$$= 1$$

$$dV = (\qquad\qquad) \; \mathbf{dr}\,\mathbf{d\theta}\,d\varphi$$

10) Let the ground state wave function of the hydrogen atom be given by:

U = exp (-ra)

Assume there is no relative motion of the nucleus so that $\theta = 0$ and $\varphi = 0$

a) Show that the relevant Schrodinger equation would now be:

$- \hbar^2 / 2m [1/ r^2 \, \partial/ \partial r \, (r^2 \, (\partial / \partial r)]U = [E - V(r))]U$

b) If $V(r) = - e^2 / r$ for the hydrogen atom, show that the quantized energy for this system is:

$E = - \hbar^2 / 2m [a^2 - 2a/ r] - e^2 / r$

c) Find the *specific energy* in the ground state if:

$r = \hbar^2 / m e^2$

and: $a = m e^2 / \hbar^2$

11) The Hamiltonian operator for the quantum harmonic oscillator is:

$\mathbf{H^\wedge} = \mathbf{p^\wedge}^2/ 2m + m\omega^2 x^2/ 2.$

Also: $\mathbf{p^\wedge} = i \hbar \, (\partial / \partial x)$. Use these to write out the applicable Schrodinger equation then indicate the form of the expected solutions – with a diagram for the potential.

Chapter V: Introduction to Nuclear Physics

5.1. Nuclear Models: The Liquid Drop Model

Nuclear models occur under a set of hypotheses, each of which explains some aspect of nuclear behavior. The most basic is perhaps the "liquid drop model" which is used to account for nuclear fission, radioactivity. This is illustrated in Fig. 1.

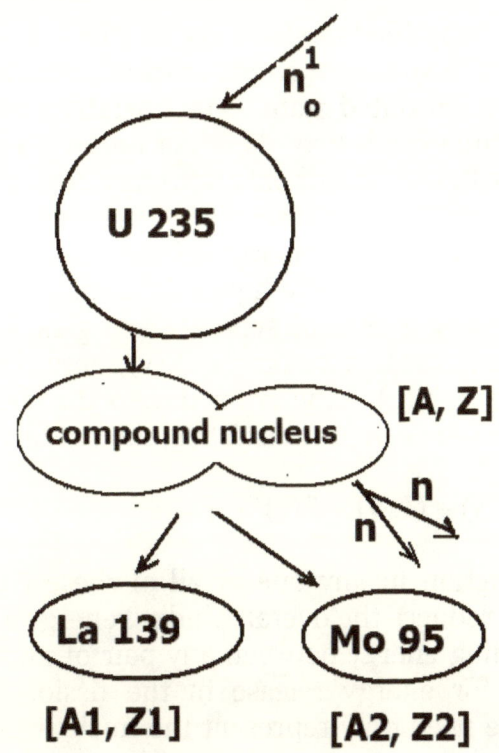

Fig. 5.1: U235 Fission in 'liquid drop' model

As the name implies, the nucleus is depicted in terms of a "liquid drop", which analogous to a liquid macroscopic drop, exhibits surface tension and an excitation energy when disturbed.

This model is premised on the exceedingly short range of nuclear forces which requires that nearest neighbor attractions predominate. This is somewhat similar to the type of attraction between the molecules of a liquid which leads to the property of its surface tension. Thus, in a liquid drop nucleus each nucleon shares its total binding energy with every nucleon.

As the simplified model diagram (for U 235) shows, we expect *a spherical configuration* in the minimum energy or unexcited state. The unstable (compound) nucleus then yields two "daughter nuclei" : La 139 and Mo 95, with two additional neutrons released

The total binding energy initially holding all the nucleons together in the liquid drop nucleus model is comprised of a sum of separate energies, including: surface or E(s), volume or E(v), and Coulombic repulsion or E(c) – the latter due to the electrostatic repulsion between protons. Thus, the binding energy:

$$E(B) = E(s) + (E(v) + E(c)$$

A reduction in any one or all of these constituent energies reduces the overall binding energy and hence the bonding energy between any pair of nucleons. To account for energy release in the fission of heavy nuclei we need only represent the nucleus as a liquid drop and its transition from the unexcited to the excited state as shown in Fig. 1.

Thus, between the spherical and compound nucleus phase, the "drop" may be said to undergo oscillations which lead it to overshoot sphericity in two directions – vertically (converting to an elliptical configuration with semi-major axis in the vertical sense, and horizontally, with semi-major axis in the horizontal sense. It finally reaches a point – the compound nucleus phase – at which the surface tension restoring force is no longer able to contain the long –range Coulombic repulsion force and the inertia of nuclear matter.

At this point, the drop reaches a stage of distortion in which the nucleus splits into two fission fragments, denoted by $[A_1, Z_1]$ and $[A_2, Z_2]$. In order to further assess the nuclear changes, we use what is called a "semi-empirical mass formula" which sets out terms as follows:

(1) $f_0(Z, A) = 1.008142Z + 1.008982 (A – Z)$

Where the Z coefficient is the mass of the hydrogen atom in atomic mass units and the $(A – Z)$ coefficient is the mass of the neutron in amu. The remaining terms listed correct for several effects which contribute to the total nuclear binding energy:

(2) $f_1 (Z, A) = -a_1 A$

This term accounts for the binding energy and is essentially proportional to the nuclear mass or the volume of the nucleus.

(3) $f_2 (Z, A) = +a_2 A^{2/3}$

This term represents a positive correction to the surface area of the nucleus, i.e. the effect of surface tension energy.

(4) $f_3 (Z, A) = + a_3 (Z^2/ A^{1/3})$

This term accounts for the positive Coulomb energy of the charged nucleus, assumed to be a sphere of radius proportional to $A^{1/3}$.

(5) $f_4 (Z, A) = + a_4 (Z - A/2)^2/ A$

This term and the one following introduce properties specific to the nucleus. This term is zero for the case of $Z = (A - Z)$ or $2Z = A$.

(6) $f_5 (Z, A) = 0, - f(A), +f(A)$

Where 0 is the result when 'Z even' pairs with $(A - Z)$ odd, or 'Z odd' pairs with $(A - Z)$ even; and $-f(A)$ is the result when Z even pairs with $(A - Z)$ even, and $f(A)$ is the results when Z odd is paired to $(A - Z)$ odd. The form of $f(A)$ is determined by fitting the data. It is found that the best fit for a simple power law relation is:

$f (A) = a_5 A^{-1/2}$

Combining all the terms we get:

$M(Z, A) =$

$1.008142Z + 1.008982 (A - Z) - a_1 A + a_2 A^{2/3}$

$+ a_3 (Z^2/ A^{1/3}) + a_4 (Z - A/2)^2/ A + (-1, 0, 1) a_5 A^{-1/2}$

The parameter values designated are: $a_1 = 0.01692$, $a_2 = 0.01912$, $a_3 = 0.000763$, $a_4 = 0.10178$, and $a_5 = 0.012$.

Worked Problem:

Evaluate the terms of the semi-empirical mass formula for the U 238 nucleus, if A = 238 and Z = 92.

Solution:

First term:

$$f_0(Z, A) = 1.008142Z + 1.008982 (A - Z)$$

$$= 1.008142(92) + 1.008982 (238 - 92) = 240.060$$

Second term:

$$f_1 (Z, A) = -a_1 A = - (0.01692) (238) = -4.02696$$

Third term:

$$f_2 (Z, A) = +a_2 A^{2/3} = (0.01912) (238)^{2/3} = 0.734298$$

Fourth term:

$$f_3 (Z, A) = + a_3 (Z^2/ A^{1/3}) = (0.00763)[(92)^2/(238)^{1/3}]$$
$$=$$
$$1.04209$$

Fifth term:

$$f_4 (Z, A) = + a_4 (Z - A/2)^2/ A = (0.010178) [92 - 119]^2/ 238$$

$$= 0.311754$$

Sixth term (e.g. $-f(A)$ since A = 238 is even, and (A − Z) = (238 − 92) is even):

$-a_5 A^{-1/2} = 0.012 (238)^{-1/2} = 7.778 \times 10^{-4}$

Summing these terms up yields: 238.12000 u, but we note the mass of the constituents by the regular mass addition formula for nuclei is:

92 (1.008142) + (238 − 92) [1.008982] = 240. 060436 u

Leading to a mass deficiency of:

ΔM = 240. 060436 − 238.12000 = 1.94 u

The binding energy is then: $E_b = \Delta M c^2 = (931$ MeV/u)(1.94) = 1806.1 MeV

Note here that MeV like eV denotes an energy which is equivalent to 1 million electron volts or:

1 MeV = 10^6 (1.6 x 10^{-19} J) = 1.6 x 10^{-13} J

From this, the **binding energy per nucleon** can also be obtained:

E_b / A = 1806.1 MeV/ 238 = 7.58 MeV / nucleon

5.2. The Nuclear Shell Model:

The shell model of the nucleus treats a different aspect, namely specific energy levels within the nucleus. To achieve this the nucleons are treated as independent particles, each of which moves in its own spherically symmetrical potential well about 50 MeV deep with rounded edges.

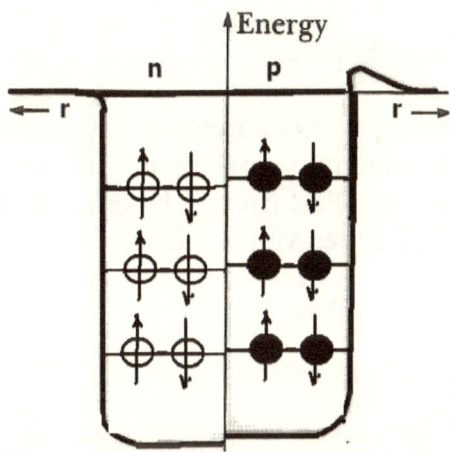

Fig. 5.2: The potential well for the Shell model

The *Schrodinger Equation* is then set up for this potential and solved. The solution yields stationary quantum states, somewhat like the states which are associated with electrons in the outer shells of the atom.

Like the electrons occupying atomic shells, the nucleons are fermions, i.e. particles each with spin ½ and therefore obeying the Pauli Exclusion Principle so that no more than two nucleons of the same type can occupy the same energy level (e.g. one with spin +½, the other with spin -½.). This leads to particles pairing up in the nucleus: spin up (+ ½) protons with spin down (-½) protons, and spin up neutrons with spin down neutrons.

The form or expression for the energy of a nucleus is very closely approximated by the energy associated with a "square well" potential. The Fermi energy in more explicit form (which we will not elaborate upon too much beyond this) is:

$$E_F = 3^{2/3} \pi^{4/3} \hbar^2 \rho^{2/3} / 2M$$

Where M is the nucleon mass, $\hbar = h/2\pi$ is the adjusted Planck number (h divided by 2π) and ρ is the nucleon density. Performing the operation with known or estimated values, the result is $E_F = 30$ MeV.

If we assume a spherical nucleus then the radius is just:

$$R = r_o A^{1/3}$$

This is also the dimension on either side of the central symmetry line of Fig. 6. The evidence for the shell model includes the following:

i)Particularly stable nuclei are borne out by "closed shells", based on our observations.

ii) The model predicts (correctly) that the even N (or A − Z) nuclei will be most stable and the odd Z, odd N nuclei the least stable. Experiments bear this out: there are 160 stable nuclides (with even Z, N) and only 4 with odd Z, N.

iii) The model predicts that for even N, Z nuclides the total angular momentum J = 0, and that for odd nuclides it is half-integral. (Borne out by measurements of nuclear magnetic moments.)

Worked Problem (2):

Find the diameter of the oxygen (O^{16}) nucleus.
Solution:

We apply: $R = r_0 A^{1/3}$

Where: $r_0 = 1.2 \times 10^{-15}$ m

Is the Bohr radius, and A = 16. Then:

$R = (1.2 \times 10^{-15}$ m$)$ $(16)^{1/3} = 3.0238 \times 10^{-15}$ m

And the diameter D = 2R = 2(3.0238 $\times$ 10^{-15} m)

D = 6.05 $\times$ 10^{-15} m

The preceding result can also be generalized to a situation in which the ratio of radii (between two nuclei) is sought. In this case, we can write:

$R_1 / R_2 = [r_0 A_1^{1/3}] / [r_0 A_2^{1/3}]$

Or:

$R_1 / R_2 = (A_1 / A_2)^{1/3}$

Example: How much larger is a copper nucleus than an oxygen nucleus?

Let R_1 denote the radius of the copper nucleus, and R_2, oxygen's. Then: $A_1 = 64$, and $A_2 = 16$. So:

$R_1 / R_2 = (64/16)^{1/3} = 4/ 2.51 = 1.587$

Thus, the copper nucleus is about 1.587 times the size of the oxygen nucleus.

5.3. Nuclear Reactions

The energy liberated in nuclear reactions is referred to the **Q- value** or the "Q of the reaction". This is the total energy released in the reaction and is usually expressed as:

$$Q = [(M + m) - M' - m']c^2$$

Where M + m denotes the sum of masses of the reactants and M', m' denotes the masses of the products. Q is obtained by using the atomic masses revealed when the reaction has been fully written out.

Before doing this it's important to recognize the various kinds of reactions and then how Q is applied, including how to obtain the kinetic energy and the initial energy:

Decay and Nuclear Fission Reactions:

We consider first natural decay and also artificial nuclear fission reactions, e.g. produced by bombardment of a nucleus by a smaller one. There are basically two types of decay processes we will look at:

Alpha decay: $_z X^A \rightarrow {}_{z\text{-}2} X^{A\text{-}4} + {}_2 He^4$

Beta Decay: $_z X^A + {}_{\text{-}1} e^o \rightarrow {}_{z\text{-}1} X^A + v$

Worked Problem (3):

Find the Q-Value of the alpha decay:

$_{84} Po^{210} \rightarrow {}_{82} Pb^{206} + {}_2 He^4$

Solution:

We find (from a table of atomic mass units) the mass of $_{84}Po^{210} = 209.982u$, the mass of $_{82}Po^{206} = 205.969u$, and $_{2}He^{4} = 4.002u$.

We confirm that typical for α-decay, the atomic weight A decreases by 4, and the atomic number Z by 2. Then we may write for the Q of the reaction:

Q = [(total rest mass before decay) −
                    (total rest mass after decay)] c^2

Q = [(209.982u) − (205.969u + 4.002u)]c^2

Q = [(209.982u) − (205.969u + 4.002u)]c^2

Q = [(209.982u) − (209. 971u)] 931 MeV/u

Q = [0.011u] 931 MeV/u = 10.24 MeV

We next seek to find the Q-value associated with the beta decay:

$_{4}Be^{7}$ + $_{-1}e^{0} \rightarrow$ $_{3}Li^{7}$ + υ

Where:

$_{4}Be^{7}$ = 9.012182 u

$_{3}Li^{7}$ = 7.016004 u

And we use: c^2 = 931.5 MeV/u

Again, using the formula for Q :

$Q = [(9.012182 \text{ u} - 7.016004] \ 931.5 \text{ MeV/u} = 1.996u$

$Q = (1.996u) \ 931.5 \text{ MeV/u} = 1859 \text{ MeV}$

Compressed notation for decays:

An abbreviated, concise notation is often used to represent nuclear reactions. Consider the case of the beta decay just analyzed. We may write for this, in concise form:

$_4\text{Be}^7 \ (_{-1}e^0 \ , \ \upsilon) \ _3\text{Li}^7$

But note that this is more often employed for the bombardment of a particle than for simple decays. For example, in the illustration above the Beryllium isn't being bombarded by anything – rather it is emitting an electron! (Though we can still use the concise notation to represent this so long as we understand the electron is a decay particle).

In general, for bombardment, given a target nucleus X bombarded by a particle a, yielding a nucleus Y and another particle b, e.g.

$a + X \rightarrow Y + b$

we have in more concise form:

$X \ (a,b) \ Y$

The Q-value of the reaction can then be worked out on the basis of:

$Q = [M_a + M_X - M_Y - M_b]c^2$

Worked Problem (4):

Write out the nuclear reaction for:

$_3Li^7 (p, \alpha)\ _2He^4$

And obtain the Q-value.

We know the p denotes the proton of hydrogen nucleus and α is an alpha particle given off. We list the nuclear masses as follows:

$_3Li^7 = 7.016004$ u

$p = _1H^1 = 1.007825$ u

$\alpha = _2He^4 = 4.002603$u

Then:

$Q = [M_a + M_X - M_Y - M_b]c^2$

$= [7.016004$ u $+ 1.007825$ u $- 4.002603$u $-$
4.002603u$]\ c^2$

$= [8.023829$ u $- 8.005206$ u$]\ 931.5$ MeV/u

So: $Q = (0.018623$ u$)\ 931.5$ MeV/u $=17.3$ MeV

5.4. Nuclear Fusion Reactions:

In general a nuclear fusion reaction is one in which two light nuclei combine (fuse) to form a heavier nucleus with positive energy given off (the Q of the reaction). Nuclear fusion is demonstrated in its most compelling form in the case of stellar energy. Exhaustive investigations in this regard, eventually led to the realization that fusion was the only practical energy by which stars could be sustained over long periods of time, such as billions of years.

In the Sun, for example, two distinct nuclear fusion processes occur: 1) the proton-proton cycle, and 2) the carbon-nitrogen cycle.

In the first of these (the easier one because it has fewer reactions):

$$^1H + {}^1H + e^- \rightarrow {}^2H + \nu + 1.44 \text{ MeV}$$

$$^2D + {}^1H \rightarrow {}^3He + \gamma + 5.49 \text{ MeV}$$

$$^3He + {}^3He \rightarrow {}^4He + {}^1H + {}^1H + 12.85 \text{ MeV}$$

The top line shows two protons fusing to yield deuterium (heavy hydrogen) with a positron and neutrino (ν) emitted, along with 1.44 MeV of energy. Empirical evidence of this reaction is obtained from gallium detectors, of the neutrinos given off, which are within 1-2% of what theoretical models predict.[4] In the second fusion reaction, the deuterium combines with a proton to give the isotope helium 3, along with a gamma ray (γ) and 5.49 MeV energy. In the final fusion, two helium-3 nuclei combine to yield one helium-4 nucleus, along with two protons, and 12. 85 MeV energy. Note that the two ending product protons

---

[4] See, e.g. **_Physics Today_**: _Reports_, April, 1995, p. 19.

commence the cycle anew, so that the generation of nuclear energy is ongoing.

The ending quantities on the right sides of each part of the cycle denote the Q of the reaction for that part. Let us check the Q for the first and simplest part. We know the hydrogen mass = *1.007825 u and for deuterium we have (from atomic tables):*
^2D = = *2.01410 u. Then:*

$$Q = [\ 2(1.007825\ u) - 2.01410\ u]\ c^2$$

$$Q = [\ 2.01565\ u - 2.015941u]\ c^2$$

$$Q = [2.01565 - 2.01410]\ 931.5\ MeV/u$$

$$Q = [0.00155]\ 931.5\ MeV/u = \textbf{\textit{1.44 MeV}}$$

The effect of ongoing fusion reactions such as this, means that the central core of the Sun becomes heavier and heavier, as more and more helium is produced. This despite the fact that the Sun as a whole is losing an amount of mass of roughly 4×10^6 metric tones per second

Inquiry Problem:

If the atomic mass for helium 3 (^{3}**He**) is equal to 3.01603 u, then verify the other Q-values for the last two parts of the proton-proton cycle. A simplified, compressed "net reaction":

^{1}H + ^{1}H +^{1}H + ^{1}H $\rightarrow$ ^{4}He + Energy

Is sometimes used to evaluate the total energy released in the proton-proton cycle. Compute this energy and compare to the value obtained for the total energy released in the earlier example. Can you account for the difference?

Nuclear Fusion Reactions in the Aging Sun:

At some stage, when nearly the entire solar core is helium a new helium fusion phase will be ushered in (at higher temperature), such that the following reaction series, known as the 'triple alpha' process, kicks in:

$$^4He + {}^4He \rightarrow {}^8Be + \gamma \ (- 95 \ keV)$$

$$^8Be + {}^4He \rightarrow {}^{12}C + \gamma + 7.4 \ MeV$$

Here, the two alpha particles (helium nuclei) first fuse to give unstable beryllium and a gamma ray (γ), with 95 keV energy *absorbed*. Then the beryllium fuses with a helium-4 to give carbon–12 plus a gamma ray and 7.4 MeV energy given off.

In this way a new cycle commences, leading to a heavier molecular weight core. Each successive burning phase, however, is less efficient than its predecessor, as can be seen by comparing the energy given off in the triple alpha process to the energy given off in the proton-proton cycle. The key thing to bear in mind in terms of a stable phase (i.e. 'Main sequence') star like the Sun is that it is in pressure-gravity balance. The outer gas pressure balances the weight of its overlying layers. Any condition likely to disrupt this balance is therefore of paramount interest.

The stable lifetime of the Sun depends on how long before it consumes ninety percent of the hydrogen in its core. Theoretical investigations using data from nuclear reaction rates and cross sections suggest the Sun's Main Sequence lifetime at 8-10 billion years. Since it already has spent 4.5 billion of those years, there are anywhere from 3.5 to 5.5 billion years remaining.

Once the triple-alpha process gets underway and the energy balance declines, the Sun will have to compensate for the lost energy to sustain any kind of balance. Thus, the Sun's core must contract and convert gravitational potential energy into thermal energy. Meanwhile, ignition of hydrogen burning in the Sun's outer layers will create radiation pressure that forces the outer layers to expands. The Sun will then become a *"Red Giant"* and its new larger surface will be expected to engulf all the planets up to and including Mars.

Worked Problem (5):

If the atomic mass of beryllium 8 (^{8}Be) = 8.00531u, verify that the first part of the triple-alpha fusion process is endothermic and has the value given.

Solution:

We have:

$Q = [\ 2(4.00260\ u) - 8.00531\ u]\ c^2$

$Q = [\ -\ 0.00011]\ 931.5\ MeV = 0.102465\ MeV = -\ 102.4\ keV$

Of course, not taken into account here is the gamma ray (γ) which also comes off. Hence we will have:

(-102.4 keV) + (E (γ)) = **-95.7 keV**

So that:

E (γ) = hc/ λ = 6.7 keV

Is the missing energy of the gamma ray photon, with the difference factored in yielding 95.7 keV.

The Problem of the Coulomb Barrier in Solar Fusion

 The problem of the Coulomb repulsive barrier to solar nuclear fusion was first highlighted and explored by Prof. Martin Schwarzschild in his excellent monograph *'The Structure and Evolution' of the Stars'* (Dover, 1958).

   Schwarzschild once calculated that the probability of any one proton fusion in the Sun's core would ordinarily be about **once every 14 billion (14 x 10⁹) years**. Since the universe itself is only 13.8 billion years old this means it could never occur unless another factor was present to enable it.

   The reason for this has to do with the Coulomb (electrostatic) repulsion between the potentially fusing H-nuclei. Thus, each proton, having (+) charge tends to repel any other proton within a discrete sphere or distance around it. (Recall from your basic physics, like charges repel, unlike attract - and that's what essentially obtains here)

   In order for thermonuclear fusion to be realized, the Coulomb barrier must be overcome. Fortunately quantum mechanics allows for a certain nonvanishing probability that a particle (say proton) of kinetic energy K, can overcome a barrier of energy V ("*barrier potential*"), via the process of "**quantum tunneling**". Note that tunneling is a general feature of low mass systems, such as single proton (H) states.

Consider a deBroglie (matter) wave arising from a single proton (p+) of form:

$U(x) \sim \sin(kx)$

Where x is the particle's linear displacement (e.g. in 1-D) and k, the wave number vector($k = 2\pi/\lambda$), where λ denotes the wavelength.

Though the associated kinetic energy **K < V** (the barrier "height") the wavefunction is *non-zero* within the barrier, e.g.

$U(x_b) \sim \exp(-cx)$

So, visualizing this behavior as shown below:

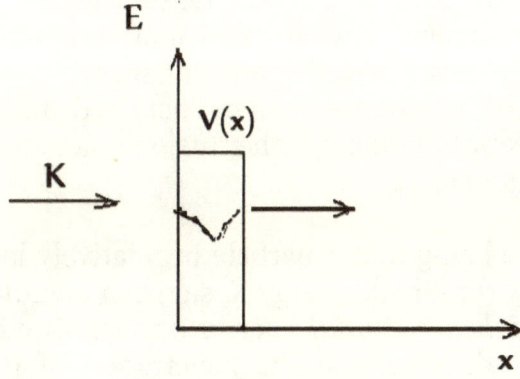

Fig. 5.3: Tunneling through Coulomb barrier

with the "barrier" at height V, so we can visualize the particle of lesser energy K, moving from the left side of the E-axis *"tunneling"* through to the right side where it may have wave function, U(x) ~sin (kx + φ), where φ denotes a phase angle.

Note that if the barrier is not too much higher than the incident particle energy, and if the mass is small, then tunneling is significant.

It's important here to point out that the penetration of the barrier is a direct result of the wave nature of matter. In effect, this wave nature - which is uniquely quantum mechanical in origin- allows a higher energy barrier to be penetrated by a lower energy particle, something totally without parallel in classical, Newtonian physics!

Even given tunneling, an "offset" is required to reduce the low penetration probability , since clearly the Sun and other stars are shining by fusion.

This 'offset' arrives via enormously high density of protons, e.g. in the core, which: i) increases the probability enormously, since so many more protons are in extremely close proximity, and enhances temperatures to the point they can be sustained, and continue - thereby building up other fusion reactions to finish the initial one.

The idea here being that a particle of relatively low incident energy (of kinetic energy K, say) can *actually penetrate a higher potential energy barrier*, say of energy V(x) > K. Note that the penetration of the barrier is a direct result of the wave nature of matter!

(The matter wave form changes in the process of transmission through the barrier, say from an exp(-ikx) function to a **sin (kx + φ)** where φ denotes phase angle). In effect, this wave nature - which is uniquely quantum mechanical in origin- allows a higher energy barrier to be penetrated by a lower energy particle, something totally without parallel in classical, Newtonian physics! Note that if the barrier is not too much higher than the incident energy, and if the mass is small, then tunneling is significant. It was insights such as this that paved the way to apprehending how much subtler nature was than hitherto realized, and how many more technological advances could be achieved when the wave nature of matter was factored into designs.

5.5. The Quantum Treatment of the Deuteron.

The deuteron is perhaps the most basic nuclear system to confront. As we know the deuteron consists of one proton, one neutron and the electron – with the first two comprising the nucleons.

To proceed, we write the usual Schrodinger equation for the hydrogen atom but let Φ and Θ = const. so that their derivatives are zero. Then, substitute in the reduced mass:

$$m' = m_n m_p / (m_n + m_p)$$

So we obtain the new form of the Schrodinger equation:

$$1/r^2 \, d/dr \, (r^2 \, dR/dr) + 2m'/\hbar^2 \, [E - V] \, R = 0$$

This can be further simplified by letting:

$U(r) = rR(r)$ which results in the equation:

$$d^2U/dr^2 + 2m'/\hbar^2 [E - V] U = 0$$

In this equation we find that V, the potential, has two values:

$V = -V_0$ and $V = 0$ outside the well. The diagram below shows how we are treating the deuteron in terms of the function V(r).

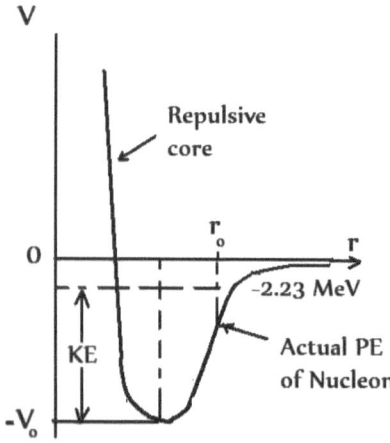

Fig. 5.4: The potential well for the deuteron

There are then two solutions we can designate:

i) u_I for $r \leq r_0$ and

ii) u_{II} for $r \geq r_0$

Inside the potential well:

$d^2 u_I / dr^2 + 2m'/\hbar^2 [E + V_o] u_I = 0$

and we let: $a^2 = 2m'/\hbar^2 [E + V_o]$

so that:

$d^2 u_I / dr^2 + a^2 u_I = 0$

For which we can show:

$u_I = A \cos(ar) + B \sin(ar)$

Since $R = u/r$ the cosine solution must be discarded, lest we get an unwanted infinity. This leaves:

$u_I = B \sin(ar)$

Outside the potential well $V = 0$ so that:

$d^2 u_{II} / dr^2 + 2m'/\hbar^2 [E] u_{II} = 0$

Let:

$b^2 = 2m'/\hbar^2 [-E]$

Since the total energy of the neutron is negative, i.e. being bound to the proton. Then:

$d^2 u_{II} / dr^2 - b^2 u_{II} = 0$

Which can be shown to have the solution:

$u_{II} = C \exp(-br) + D (\exp(br)$

For consistency we demand $u \to 0$ as $r \to \infty$,

So D = 0 and:

$u_{II} = C \exp(-br)$

For continuity at $r = r_0$ and $u_I = u_{II}$:

$B \sin(ar) = C \exp(-br)$

Thence:

$du_I/dr = du_{II}/dr \Rightarrow aB \cos(a r_0) = -bC e^{-br_0}$

Now, divide the solution on the left side by the one on the right side, e.g.:

$A B \cos(a r_0) / -b C e^{-br_0}$

so:

$\tan(a r_0) = - a/b$

In effect, the deuteron problem cannot be solve analytically only graphically.

Quantum Numbers for Deuteron:

Since there are two particles each with intrinsic spin $\frac{1}{2}$ in the deuteron, the total intrinsic spin angular momentum can only have the values: $S = s_1 \pm s_2 = 0$ or 1. The orbital angular momentum quantum number, L (describing the motion in space of the proton and neutron relative to each other) can assume the values L= 0, 1, 2 (i.e. S, P, and D states).

The total angular momentum of the deuteron has been measured and the total angular momentum quantum number J has been found to be 1. This must be the vector sum of the orbital angular momentum (L) and the total spin momentum (S).

End of Chapter Problems:

1) Evaluate the terms of the semi-empirical mass formula for the U 238 nucleus, if A = 238 and Z = 92. Use this information to find the total mass in atomic mass units (u) and compare it to the standard mass expression.

2)The sketch graph shown below plots the mass number A vs. the binding energy per nucleon (BE/nucleon) on the vertical axis.

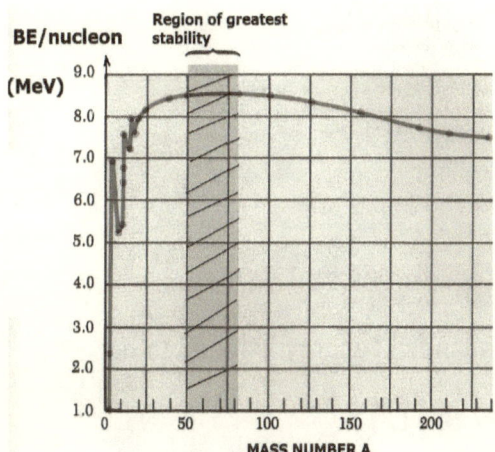

Fig. 5.5: The nuclear binding energy as a function of mass number A

Account for, using any models or other explanations, the region of greatest stability indicated.

3) A quantum square well potential is defined according to:

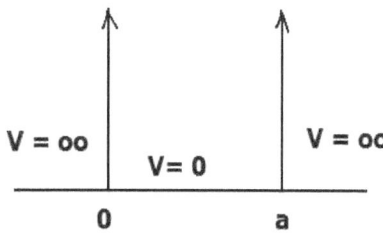

Infinite square well quantum system. Particle must be between 0 and a!

It is found that the Schrodinger equation to solve becomes:

$-\hbar^2/\, 2m\, (\, d\psi^2/dx^2\,) = E\psi$

Thence:

$d\psi^2/dx^2 + 2m/\,\hbar^2\, (E\psi) = 0$

And the quantized energy is found to be:

$E_n = (h^2/\, 8m\, L^2)\, n^2$

This can be modified to yield a simplified *Fermi energy* in one dimension by using instead:

$E_n = (h^2 \pi^2/ 8m L^2)(N/2)^2$

Where all the Fermi shell energy levels are presumed occupied up to $N/2$. Use this to obtain a simplified estimate of the energy associated with the oxygen **nucleus** if we assume its shells filled up to $N = 4$ and use an estimate for L as $R = r_o A^{1/3}$

(Take $m = 1.7 \times 10^{-27}$ kg)

4)(a) Compare the value you obtain in (3) to the actual Fermi (shell) energy for oxygen, if the nuclear density $\rho \approx 2.3 \times 10^{17}$ kgm^{-3}.

(b) Write out at least one technique that might be used to find the density of an atomic nucleus. State clearly any assumptions made and how you quantify them.

5)(a) Find the ratio of the helium nucleus to the uranium 238 nucleus.

(b) Estimate, using any technique you can think of, the ratio of the nuclear densities for part (a).

6) An element has mass number $A = 202$ and atomic number $Z = 80$.

a) Find the diameter of the nucleus and how many times it is greater than that of hydrogen.

b) Find the mass defect ΔM for this nucleus.

From the result in (b) obtain the binding energy and the binding energy per nucleon, E_B/A.

7) Calculate the wavelength of the gamma ray photon (in nm) which would be needed to balance the endothermic part of the triple −alpha fusion equation. (Recall here that 1 eV = 1.6 x 10 $^{-19}$ J)

8) Verify the second part of the triple-alpha fusion reaction, especially the Q-value. Account for any differences in energy released by reference to the gamma ray photon coming off and specifically, give the wavelength of this photon required to validate the Q.

9) The luminosity or power of the Sun is measured to be L = 3.9 x 10^{26} watts. Use this to estimate the mass (in kilograms) of the Sun that is converted into energy every second. State any assumptions made and reasoning.

10) The half −life of U238 is 4.5 x 10^9 yr. and the half life of U235 is 7.1 x 10^8 yr. These isotopes occur in nature in the ratio:

U 238 atoms/ U235 atoms = 140

If Earth is 4.0 x 10^9 yr. old what was the ratio at the time of Earth's formation?

11) An alpha-unstable nucleus doesn't disintegrate immediately because of the Coulomb barrier. Explain from the shape of this barrier why those isotopes which emit lower energy α-particles have longer half lives.

12) Estimate the height of the Coulomb barrier for an α-particle in the vicinity of the gold (Au) nucleus.

13) When ^{118}Sn$_{50}$ is bombarded with a proton the main fission fragments are: ^{24}Na$_{11}$ and ^{94}Zr$_{40}$

The excitation energy necessary for passage over the potential barrier is:

$\varepsilon > 3 k e^2 Z^2 / 5$

Where the right hand side denotes the height of the Coulomb barrier.

a) What must this value be?

(Take $ke^2 = 1.44$ MeV/ fm)

14) Find *the energy difference* between the reactants and the products. (Take $c^2 = 931.5$ MeV/ u)

15) Show that $u_I = A \cos(ar) + B \sin(ar)$

Is a solution of the reduced Schrodinger equation for the deuteron, $d^2 u_I / dr^2 + a^2 u_I = 0$.

16) Show why the cosine solution needs to be discarded. Show that $u_{II} = C \exp(-br)$ is a solution of the other reduced Schrodinger equation for the deuteron:

$d^2 u_{II} / dr^2 + b^2 u_{II} = 0$.

17) Show by using the quantum numbers J, L and S, that the deuteron is describable *by only 4 states*: $^3S_1$, $^3D_1$, $^1P_1$ and $^3P_1$ (Hint: Use L-S coupling techniques.)

Chapter VI. Quantum Statistical Physics

6.1. The Occupancy of States and Distributions:

In statistical physics two particular distributions occupy attention: the Bose-Einstein, and the Fermi-Dirac. For our purposes we will spend much more time on the latter but it is useful to see how both enter the picture.

In the case of the Bose-Einstein formalism we are concerned with bosons which have integral spin values (i.e. an orbital may be populated by any number of bosons so the Pauli Exclusion principle doesn't apply).

Consider the distribution function for a system of non-interacting bosons in which the system is in thermal and diffusive contact with a reservoir. Thus we may allow a circumstance such as this:

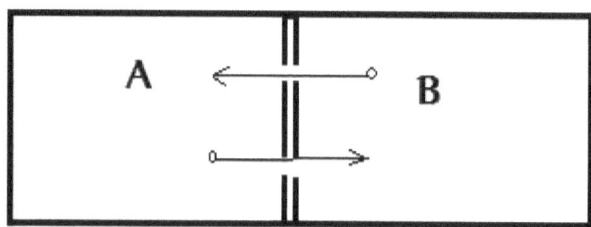

Fig.6. 1: Illustrating systems in thermal and diffusive contact.

Thus heat can be exchanged between the systems A and B, as well as particles. In the formalism treatment we let ε denote the energy of a single orbital when occupied by one particle. Hence, when n particles occupy the orbital then the energy must be n ε. To simplify, we treat one orbital as a system and ignore all others. The grand sum taken over one orbital would then be:

$$\sum^{\infty}_{n=0} \sum_{e} \exp(N\mu - \varepsilon)/\tau) = \sum^{\infty}_{n=0} \lambda^n e^{-n\varepsilon/\tau}$$

$$= \sum^{\infty}_{n=0} (\lambda e^{-n\varepsilon/\tau})^n$$

Now, let $x = \lambda e^{-n\varepsilon/\tau}$ and sum the series in closed form so the grand sum is now:

$$\mathcal{Z} = \sum^{\infty}_{n=0} (x)^n = 1/1-x = 1/1-\lambda e^{-n\varepsilon/\tau}$$

Provided: $\lambda e^{-n\varepsilon/\tau} < 1$

The ensemble average of the number of particles in the orbital is by definition of the average value:

$$< n(\varepsilon) > = \sum^{\infty}_{n=0} (nx)^n / \sum^{\infty}_{n=0} (x)^n =$$

$$x(d/dx) \sum^{\infty}_{n=0} (x)^n / \sum^{\infty}_{n=0} (x)^n =$$

$$x(d/dx)(1-x)/(1-x)^{-1}$$

After some calculus and messy algebra we find:

$$< n(\varepsilon) > = 1/\exp(\varepsilon - \mu)/\tau$$

Which is the Bose-Einstein distribution.

For the Fermi-Dirac distribution we consider a simplified system represented by a cubical box of volume $V = L^3$

The number of electrons (Fermi particles) within occurs between energies ε and $\varepsilon + d\varepsilon$. The energies occur in a quantized way and are depicted schematically below:

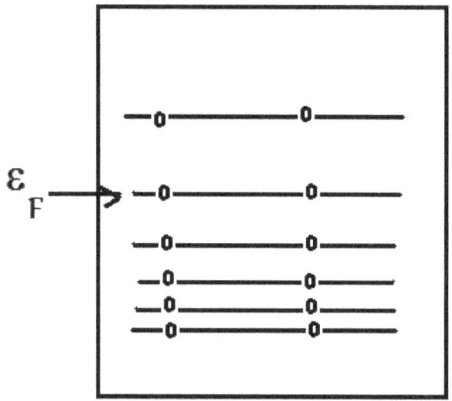

Fig 6.2. Box with Fermi occupancy and energy levels.

The diagram shows each level corresponding to two orbitals, one for spin up the other for spin down. The Fermi energy:

$$\varepsilon_F = \hbar^2 \, (\pi \, n_F / L)^2 / 2m$$

Is determined by the requirement that the system in the ground state hold N electrons – with each orbital filled with one electron if the energy of the orbital is less than: ε_F.

Since for a 3D box the quantized energy is:

$\hbar^2/2m \, [\pi \, n^2 / L]$

where $n^2 = n_x^2 + n_y^2 + n_z^2$

Then the number of states possessed by a 3D box is:

$1/8 \, [4\pi \, n^3 / 3]$

For a sphere, the number of orbitals in some radius is:

$n = \gamma \, [4\pi \, n^3 / 3] \, (1/8) = \gamma \pi \, n^3 / 6$

For electrons, $\gamma = 2$ so:

$n = \pi \, n^3 / 3$

Hence, if the box holds N electrons the orbitals must be filled up to the quantum number n_F. So the total number of electrons is:

$N = \pi \, n_F^3 / 3$ so $n_F = (3N/\pi)^{1/3}$

The Fermi energy would then be:

$\varepsilon_F = \hbar^2 \, (\pi / L)^2 \, n_F^2 / 2m =$

$\hbar^2 \, (\pi / L)^2 \, (3N/\pi)^{1/3} / 2m$

Since $V = L^3$ we can write:

$$\varepsilon_F = \hbar^2 / 2m \ [3\pi^2 N/v]^{2/3}$$

This yields good results for monovalent atoms because they have only one electron per atom.

6.2. Treating the Nucleus as a Fermi Gas

One of the useful applications of the Fermi quantum statistics is to the atomic nucleus, such as that for the deuteron. The key to the application is assuming an independent particle model so each nucleon moves in a smooth potential hole as opposed to being subject to the actions of (A - 1) other nucleons. In this sense, the model would exhibit similar properties to the "electron gas" which we examined being confined to a 3D box in the previous section. In this case, the nucleons will be constrained to move in a spherical potential hole of radius:

$$R = r_o A^{1/3}$$

(I.e. analogous to the case of electrons)

We focus then on all momentum states being filled up to the Fermi momentum, p_F.

Since:

$$p_F^2 / 2m = \hbar^2 / 2m \ [3\pi^2 N/v]^{2/3}$$

the number of states up to p_F per unit volume is:

$$N/v = 4\pi/3 \ (p_F^3/h^3)$$

Or the volume in momentum space divided by the Planck constant. Let's now apply this to the case of the deuteron, for which we have a neutron and proton in each state and each can have spin up or spin down. Then we get a total of four states so multiply the expression by 4 to get:

$$4N = 4[4\pi/3 \, (c\,h^3) \, (4\pi R^3/2\pi h^3) \, 3]$$

Simplifying:

$$p_F r_o/h = \sqrt{(9\pi/8)} = 1.52$$

This prescribes: the maximum momentum present in the nucleus in terms of the nuclear radius r_o.

$$p_F = 1.52 \, (h/r_o)$$

Taking $r_o = 1.1$ fm (e.g. 1.1×10^{-15} m)

We learn the maximum kinetic energies in the nucleus are as large as:

$$K_F = p_F^2/2m = 39 \text{ MeV}$$

If the binding energy of the 'last nucleon', e.g. at the top of the Fermi distribution) is ≈ 8 MeV then this result shows that the nucleus looks to each nucleon like a potential hole of 39 MeV depth.

Worked Problem:

Find the Fermi sphere parameters: ε_F, v_F and T_F for He3 at absolute zero, viewed as a gas of non-interactive fermions. (The density of the liquid is 81 kg/ m^3).

Solution:

We first find the number of electrons per unit volume. (N/V)

N/ V =

6.020 x 10^{26} atoms/ kmol (81 kg/ m^3)/ 3 kg/ kmol

(Why 3 kg/ kmol in denominator? Because the atomic weight = 3)

Then: N/ V = 1.625 x 10^{28} atoms

The Fermi Energy, ε_F = h^2 / 2m [3π^2 N/ v]$^{2/3}$

Then:

ε_F = 6.81 x 10^{-23} J

To get: v$_F$, note: ε_F = ½ m (v$_F$)^2

v$_F$ = [2 ε_F / m]$^{1/2}$ = 165 m/s

Finally, the Fermi temperature is found based on the fact it is tied to the energy (as the energy increases, the temperature increases. We have for the thermodynamic temperature:

Since T$_F$ = k$_B$ τ Then: T$_F$ = (ε_F)/ k$_B$ =

(6.81 x 10^{-23} J) / (1.38 x 10^{-23} JK^{-1}) = 4.93 K

End of Chapter Problems:

1) Find the Fermi sphere parameters: e_F, v_F and T_F for He3 at absolute zero, viewed as a gas of non-interactive fermions. (The density of the liquid is 81 kg/ m³).

2) a) Show that $(- \partial f / \partial \varepsilon)$ evaluated at the Fermi level ($\varepsilon = \mu$) has the value $(4 k_B T)^{-1}$. Thus, the lower the temperature, then the steeper the slope of the Fermi-Dirac function.

Hint: Use $f(\varepsilon) = 1/ \{\exp (\mu - \varepsilon)/ \tau + 1\}$

b) Make a careful plot of $(- \partial f / \partial \varepsilon)$ vs. ε/ k_B for the specific case: $\mu/ k_B = 5 \times 10^4 K$ and $5 \times 10^2 K$.

3) Let $\varepsilon = \mu + \delta$, show that: $f(\delta) = 1 - f(-\delta)$

Hint: Let $f(\delta) = 1/ [\exp (\mu - \varepsilon)/ \tau + 1]$

4) Treating the Tritium nucleus as a nucleonic electron gas, assume an independent particle model (based on Fermi energy levels) so each nucleon moves in a smooth potential hole.

Thereby obtain the maximum momentum present in the nucleus in terms of the nuclear radius r_o.

$p_F = 1.52 (h/ r_o)$

(Take $r_o = 1.1$ fm)

Use this to estimate the depth of the potential hole, i.e. the magnitude of the well in MeV

Chapter VII. Tensors and General Relativity

7.1. Introduction to Tensors.

Consider the vector diagram below showing a mass suspended by two ropes under tensions T_A and T_B and weight W, which we refer to as vectors.

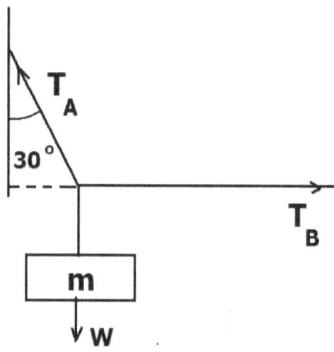

Fig. 7.1: Diagram of forces in equilibrium

The specific vector diagram one simplified is:

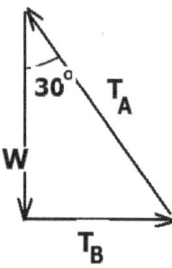

Fig. 7.2. Vectors shown for the force diagram

We say that vectors such as **W**, **T**$_A$ and **T**$_B$ are quantities that can be designated using a single index, e.g. **W** $_i$ or **T**$_{Bi}$. These may be said to transform into each other in a specified way under a rotation of axes, for example. But there are also quantities used in physics that *have more than one index* and transform into each other in more complicated ways.

Generalizing, if there are n indices we say the tensor is of rank n. Analogously, a *vector* – say **W**, **T**$_A$ or **T**$_B$ is a tensor of rank 1. A scalar then is a tensor of rank 0.

Consider a case of a tensor of rank 2, by looking at the angular momentum, L:

$$\mathbf{L} = m\,\mathbf{r}\ \mathbf{x}\,\mathbf{v}$$

$$\mathbf{L} = m\,\mathbf{r}\,x\,(\omega\ x\ \mathbf{r}) = m\,(\mathbf{r}^2\,\omega\ - (\mathbf{r}\,\cdot\,\omega)\,\mathbf{r}\,)$$

Where: $\mathbf{r}^2 = x^2 = x_{kk}$

The last equation for angular momentum can also be expressed:

$$L = m\,(x^2\,\omega\ -(x\cdot\,\omega)\ x\,)$$

Where $x^2 = x_k\,x_k$

And we've used the expression for the triple vector product:

$$\mathbf{A}\ x\,(\mathbf{B}\ x\,\mathbf{C}) = (\mathbf{A}\cdot\mathbf{C})\,\mathbf{B}\ -\ (\mathbf{A}\cdot\ \mathbf{B})\,C$$

Using components the preceding equation in L can be expressed:

$$L_i = m[x^2 \omega_i - x_j \omega_j x_i] = I_{ij} \omega_j$$

Where repeated indices are summed over, and the moment of inertia I_{ij} is given by:

$$I_{ij} = m[x^2 \delta_{ij} - x_j x_j]$$

Where δ_{ij} is the mixed tensor of rank two which arises from coordinate transformation and is called the Kronecker delta. Defined as:

$$\delta_{ij} = \{0 \text{ if } i \neq j\}$$

$$\delta_{ij} = \{1 \text{ if } i = j\}$$

Tensors are also the language of general relativity, i.e. the *space-time interval*:

$$ds^2 = \partial x_i / \partial q^j \; \partial x_i / \partial q^k \, dq^j dq^k = g_{ik} \, dq^i dq^k$$

where the superscripts denote particular contravariant operations. Then g_{ik} is a *matrix*:

$$g_{ik} =$$

$$(1.....0...............0)$$

$$(0.....r^2...............0)$$

$$(0.....0......r^2 \sin \phi)$$

Called the "metric tensor".

Further, $g_{ik} \, dq^i \, dq^k$

$=$

$(1.....0...............0)\ (dr^2)$

$(0.....r^2...............0)\ (d\theta^2)$

$(0.....0.......r^2 \sin \phi)\ (d\phi^2)$

So the operations applied to matrices can be applied to tensors.

In using tensors we take care with the subscripts and superscripts and use the first for *covariant* tensors and the second for *contra-variant* tensors.

The most basic tensor of all is the unit tensor, idemfactor or identical dyadic. defined:

$\underline{I} = i^\wedge i^\wedge + j^\wedge j^\wedge + k^\wedge k^\wedge =$

$$\begin{pmatrix} 1 & 0 & 0 \\ 0 & 1 & 0 \\ 0 & 0 & 1 \end{pmatrix}$$

When $\underline{I}$ is multiplied by any other vector as a factor the vector is unchanged, i.e.

$\underline{I} \cdot \mathbf{C} = i^\wedge \, C_x + j^\wedge \, C_y + k^\wedge \, C_z$

Further properties:

A tensor is *symmetric* if: $T_{ij} = T_{ji}$

A tensor is *anti- symmetric* if: $T_{ij} = - T_{ji}$

Every symmetric tensor will have the form: $a_{ij} =$

$(a_{11} ......a_{12}.........a_{13})$

$(a_{12}..... a_{22}..........a_{23})$

$(a_{13}......a_{23}.........a_{33})$

The anti-symmetric or skew symmetric tensor will have the form: $a_{jI} =$

$(0 ......a_{12}......-a_{31})$

$(-a_{12}.....0.........a_{23})$

$(a_{31}......- a_{23}...\ \ 0)$

Or effectively only three distinct components.

The Kronecker delta, δ_{ij} =

$$\begin{pmatrix} 1 & 0 & 0 \\ 0 & 1 & 0 \\ 0 & 0 & 1 \end{pmatrix}$$

Worked Problem: Compute: $a_{ij} \delta_{Ij}$

Solution:

If a_{ij} is a second order tensor with matrix:

$$a_{ij} := \begin{pmatrix} 3 & 0 & 1 \\ 1 & 2 & 5 \\ -1 & 4 & 2 \end{pmatrix}$$

Then we obtain:

$$\begin{pmatrix} 3 & 0 & 0 \\ 1 & 2 & 5 \\ -1 & 4 & 2 \end{pmatrix} \begin{pmatrix} 1 & 0 & 0 \\ 0 & 1 & 0 \\ 0 & 0 & 0 \end{pmatrix} \cdot = \begin{pmatrix} 3 & 0 & 0 \\ 1 & 2 & 5 \\ -1 & 4 & 2 \end{pmatrix}$$

7.2. Principal Axis – Diagonalization.

Doubled dummy indices, e.g. ii, jj, kk refer to the **trace** of a matrix, or the sum of the diagonal elements. For example, if: $i^\wedge i^\wedge + j^\wedge j^\wedge + k^\wedge k^\wedge =$

$$\begin{pmatrix} 1 & 0 & 0 \\ 0 & 1 & 0 \\ 0 & 0 & 1 \end{pmatrix}$$

Then the trace of the matrix = 1 + 1 + 1 = 3.

Diagonalizing tensors is analogous to obtaining the eigenvalues for a matrix in linear algebra. (See Mathematics Supplement) Hence, we need to extract the eigenvalue equations via diagonalization and obtain the distinct eigenvalues.

Consider the solid tetrad shown below:

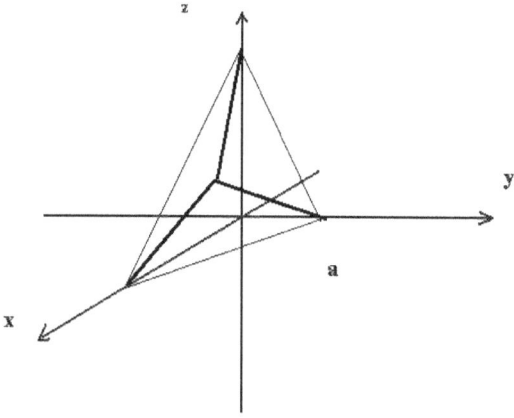

Fig. 7.3. Object for finding the principal axis

We define a_{ij} =

$$\begin{pmatrix} \dfrac{1}{\sqrt{2}} & -\dfrac{1}{\sqrt{2}} & 0 \\[2mm] \dfrac{1}{\sqrt{2}} & \dfrac{1}{\sqrt{2}} & 0 \\[2mm] 0 & 0 & 1 \end{pmatrix}$$

With $\mathbf{T'} = \mathbf{A} \cdot \underline{\mathbf{I}} \cdot \mathbf{A^t}$

Where $\mathbf{A^t}$ denotes the transpose. Then we obtain, $\mathbf{T'}$ =

$$\begin{pmatrix} 15 & 0 & 0 \\ 0 & 11 & -3\cdot\sqrt{2} \\ 0 & -3\cdot\sqrt{2} & 8 \end{pmatrix}$$

Which is to be diagonalized. Writing this out:

(15 - λ…..0……..0)

(0…….11- λ ….-3√2)

(0……..-3√2…8- λ)

leads to a cubic with triple roots which are:

$\lambda_1 = 15$, $\lambda_2 = 5$, and $\lambda_3 = 14$

Substituting λ_1 in the matrix we get:

$$\begin{pmatrix} 0 & 0 & 0 \\ 0 & -4 & -3\cdot\sqrt{2} \\ 0 & -3\cdot\sqrt{2} & -7 \end{pmatrix} \cdot \begin{pmatrix} c_x \\ c_y \\ c_z \end{pmatrix}$$

For which the separate equations, e.g. in c_x, c_y and c_z can be solved. For example,

$$-4\cdot c_y - 3\cdot\sqrt{2}\cdot c_z := 0$$

$$-3\cdot\sqrt{2}\cdot c_y - 7\cdot c_z := 0$$

After working through all the solutions, we obtain:

$$\underline{C} = -\sqrt{(2/3)}\, \mathbf{e_2}\hat{} + 1/\sqrt{3}\,(\mathbf{e_3}\hat{})$$

Worked Problem:

In a certain rectangular coordinate system, the directions of whose axes are given by the unit vectors $\mathbf{i}, \mathbf{j}$ and $\mathbf{k}$, the inertia tensor of an object is given by:

$$I = K\,x$$

$$
(1....0.....0)
$$
$$
(0....1.....1)
$$
$$
(0....1... ..1)
$$

a) What are the *principal moments of inertia* of the object (the moments of inertia along the principal axis) relative to the origin of the above coordinate system?

b) What is the direction of the principal axis corresponding to the principal moment of inertia and equal to K?

c) If the origin of the above rectangular coordinate system is at the **center of mass of the object** and the total mass of the object is M, what is the change in the inertial tensor of the object if the rectangular coordinate system is displaced parallel to itself a distance r_0 in the direction

$$(1/\sqrt{2})\mathbf{j} + (1/\sqrt{2})\mathbf{k}?$$

Solution:

a) We have: $I = K\mathbf{x}$

$$
\begin{pmatrix}
1 & \dots 0 & \dots 0 \\
0 & \dots 1 & \dots 1 \\
0 & \dots 1 & \dots 1
\end{pmatrix}
$$

We write out the determinant with eigenvalue λ:

$$
\begin{pmatrix}
1-\lambda & \dots 0 & \dots 0 \\
0 & \dots 1-\lambda & \dots 1 \\
0 & \dots 1 & \dots 1-\lambda
\end{pmatrix}
$$

Leading to the characteristic equation:

$$(1-\lambda)^3 - (1-\lambda) = 0$$

Factoring:

$$(1-\lambda)\,[\,((1-\lambda)^2 - 1\,] = 0$$

Or:

$$(1-\lambda)(\lambda^2 - 2\lambda) = 0$$

Yielding eigenvalues: $\lambda = 0,\ \lambda = 2$

Then:

$T = K\lambda$, so:

$T_1 = 0$, $T_2 = K$, and $T_3 = 2K$

Or: $(0, K, 2K)$

b) We have to take: $(I - T_1)\underline{C,}$ so that:

K [(1....0.....0)
  [(0....1.....1)
  [(0....1.....1)
-
K (1....0.....0)](x)
  (0....1..... 0)] (y)
  (0....0.....1)] (z)

=
  (0....0.....0) (x)
  (0....0.....1) (y)
  (0....1.....0) (z) = 0

Therefore: 0 =

(o)
(z)
(y)

    Where '0' for x implies the answer is **i.**

b) By the analog of the **parallel axis theorem**:

$I_o = I_G - M(\mathbf{R}^2 I - RR)$

$\Delta I = I_o - I_G = M(\mathbf{R}^2 I - RR)$

$RR = r_o^2 \ x =$

$(0....0.........0)$
$(0....1/2.....1/2)$
$(0....1/2.....1/2)$

$\Delta I = M \ r_o^2 \ x$

162

$$[(1....0.....0) \quad\quad (0....0........0)]$$
$$[(0....1.....0) \quad - \quad (0....1/2.. 1/2)]$$
$$[(0....0... ..1) \quad\quad (0....1/2... ..1/2)]$$

$$= \mathbf{M\ r_0^2\ x\ } =$$

$$(1....0.............0)$$
$$(0....1/2.....-1/2)$$
$$(0....-1/2... .1/2)$$

7.3. The Principle of Equivalence and Geodesics

Up to now, all the tensor descriptions have been for 3-space and we've not introduced time. But this deficiency must now be remedied as we provide for an extension and generalization of the Principle of Equivalence to incorporate the fact that the velocity of light is the same in all reference frames.

To put it into terms consistent with the examples at the outset of this chapter (Section 1), a mass point carried by a gravitational field does not remain at rest but "falls down". More to the point, the path taken for this freely falling mass is a geodesic in space-time or a "world line" for the mass, i.e. described in 4 dimensions.

Thus, the formalism or representation must include changes in time and so the gravitational field is now given as a **space-time tensor**. Unlike the tensors shown in the previous section, this one displays 4 x 4 components, as opposed to 3 x 3.

The reframed Principle of Equivalence therefore states that the equations of motion for this freely falling particle are expressed:

$$d^2x^k/ds^2 + \{mn, k\}\, dx^m/ds\ dx^n/ds = 0$$

(where $m, n, k = 1, 2, 3, 4$ for the four dimensions which are then displayed as: x_1, x_2, x_3 and x_4 – the last instead of t). It is important to point out that s, the interval, is assigned the dimension of time and hence is the *proper time*. By virtue of the definition and the above equation, the observer has a full definition of a curve traced out in **space-time** by a moving clock – which possesses an invariant form – and thus has a coordinate system for all events in the universe.

Let a light ray follow this path and one obtains a null geodesic, such as represented below:

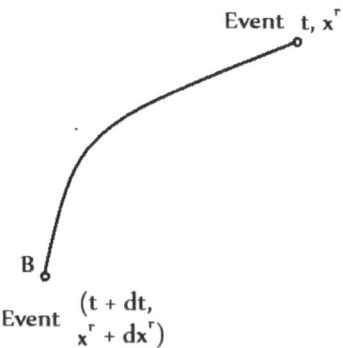

Fig. 7.4. Null Geodesic between two events

Then for a clock traveling *with the mass* (B) the proper time between the two events is zero – given the path is a null geodesic. Hence:

$$g_{44} (t, x_r) \, dt^2 = 1/ c^2 \sum_{l,m=1}^{3} g_{lm} (t, x_r) \, dx^l \, dx^m$$

However, for a clock *at rest* at position x_1, x_2, x_3 the proper time elapsed between the events shown in the diagram is:

$$ds_0 = [g_{44} (t, x_r)]^{1/2} \, dt$$

The distance between the two events exclusively in space coordinates is:

$$d\ell = \sum_{l,m=1}^{3} g_{lm} (t, x_r) \, dx^l \, dx^m$$

Then, taking the derivative: $d\ell / ds_0$

$$d\ell / ds_0 = \pm c$$

where c is called *the velocity of light* at the event (t, x_r). Hence, we see it emerges as a universal constant and always takes the *least time* between two events, hence the path taken of a null geodesic. This is consistent with the Einsteinian postulate – also given in special relativity – that the velocity of light has the same value at all points in space –time.

   Thus we see that in order to include non-uniform fields the Principle of Equivalence can be reformulated – to a "strong" form, to read[5]:

---

[5] Symon, K.R.: *'Relativistic Dynamics'*, in **Mechanics**, Addison-Wesley Publishing Co., 577, 1971.

"In a small, freely falling laboratory, the laws of physics are the same as the laws of special relativity without any gravitational field."

7.4. Einstein's Field Equations

In the previous section we saw the coefficient g_{44} appear for the expressions to do with proper time. But what is it? This denotes the gravitational potential field given by a particular set of components, usually written as $\mathbf{g}_{\mu\nu}$ for the tensor g.

Then g_{44} denotes the 16th (last) element of the $\mathbf{g}_{\mu\nu}$ matrix. In flat space-time the tensor components for g are usually given as:

$$\begin{pmatrix} g_{11} & g_{12} & g_{13} & g_{14} \\ g_{21} & -g_{22} & g_{23} & g_{24} \\ g_{31} & g_{32} & g_{33} & g_{34} \\ g_{41} & g_{42} & g_{43} & g_{44} \end{pmatrix}$$

For which it is convenient to specify special values of the potentials presented in "standard form" as:

$$\begin{matrix} g_{11} & g_{12} & g_{13} & g_{14} \\ & g_{22} & g_{23} & g_{24} \\ & & g_{33} & g_{34} \\ & & & g_{44} \end{matrix}$$

For **flat space-time** the values are all 0 except for those along the diagonal, for which:

$$g_{11} \quad = -1$$

$\mathbf{g}_{22} \quad = -1$

$\mathbf{g}_{33} \quad = \mathbf{-1}$

$\mathbf{g}_{44} \quad = \quad 1$

Thus, we see that the value of $\mathbf{g_{44}}$ in the previous expressions is 1. Writing out the interval form for the above is straightforward and one only needs to include the correct subscripts for the respective dx's, in each component, e.g.

$\mathbf{g}_{11} dx_1^2 \quad$ and $\quad \mathbf{g}_{12} dx_1 dx_2$

Now, if we take: $\partial \mathbf{g}_{\mu v} / \partial x_\tau$

We obtain the Riemann-Christoffel curvature tensor[6]:

$\Gamma^\tau{}_{\mu v} = G_{\mu v}$

For flat space-time the gravitational potentials satisfy: $G_{\mu v} = 0$

Where the values conform to those of the g's shown above and we find:

$\mathbf{g}_{11} \quad = \quad -1/\gamma$

$\mathbf{g}_{22} \quad = -x_1^2$

$\mathbf{g}_{33} \quad = - x_1^2 \sin^2 x_2^2$

$\mathbf{g}_{44} \quad = \quad \gamma$

---

[6] This is actually not a real tensor but only *a linear* tensor, i.e. vector.

Where: $\gamma = 1 - \kappa / x_1$

And the constant κ is what's called the Gaussian curvature. It can assume values of 0 (Euclidean 4-D flat space), -1 (Lobachevskian space) or +1 (Riemannian space).

Worked Problem (2): Assume a metric:

$$g_{11} = 1, \quad g_{22} = x^1, \quad g_{33} = x^2$$

Then the non-zero Christoffel symbols have values:

$$\Gamma^1{}_{22} = -1, \quad \Gamma^2{}_{12} = \Gamma^2{}_{21} = 1/x^1$$

$$\Gamma^3{}_{23} = \Gamma^3{}_{32} = -1/x^2$$

Find: a) $g_{11} \ \Gamma^2{}_{21}$ and (b) $g_{22} \Gamma^3{}_{32}$

Solutions:

a) $g_{11} \ \Gamma^2{}_{21} = (1)(1/x^1) = 1/x^1$

b) $g_{22} \Gamma^3{}_{32} = x^1 (-1/x^2) = -x^1/x^2$

The Einstein field equations can be summarized in the tensor form:

$$G_{\mu\nu} = -\tfrac{1}{2} g_{\mu\nu} G = -8 \pi T_{\mu\nu}$$

Where $\mathbf{T}_{\mu\nu}$ denotes the associated "stress-energy" tensor including internal stresses, density of matter and its component velocities u, v, w, (or in some texts: u1, u2 and u3). From this one can see that if no matter is present, one would have: $G_{\mu\nu} = 0$.

If matter is present there must be internal stresses and velocities so that: $G_{\mu\nu} = K_{\mu\nu}$ where (as seen from the field equations): $K_{\mu\nu} = -8\pi T_{\mu\nu}$

The introduction of $K_{\mu\nu}$ to describe the matter-associated properties is generally attributed to Sir Arthur Eddington[7]. We have then for the $T_{\mu\nu}$ analogous to the g's in standard form[8]:

$$
\begin{array}{cccc}
T_{11} & T_{12} & T_{13} & T_{14} \\
& T_{22} & T_{23} & T_{24} \\
& & T_{33} & T_{34} \\
& & & T_{44}
\end{array}
$$

$$
\begin{array}{llll}
= p_{11} + \rho\, u^2, & p_{12} + uv, & p_{13} + \rho\, uw, & -\rho u \\
& p_{22} + \rho\, v^2, & p_{23} + \rho\, vw, & -\rho v \\
& & p_{33} + \rho\, w^2, & -\rho w
\end{array}
$$

Which again, is a vastly simplified presentation.

It should also be said that inclusion of radiation, say via the electromagnetic energy tensor, will have the same rank as the $T_{\mu\nu}$. These components can thus be added to components of the mass energy tensor shown. In other words, the presence of

[7] Eddington, A.S.:1920, **Space, Time and Gravitation**, Cambridge University Press,, p. 173.
[8] *Op. cit.*, p. 176.

radiation is taken to be *equivalent* to the presence of mass, given m = E/ c^2.

After inserting the stress energy tensor equations into the Einstein field equations one gets:

$$(dR/dt)(/R)^2 = (8\pi)/3 (G\rho) - \kappa/R^2$$

whence:

$$(d^2R/dt^2)/R = -4\pi G_{\mu\nu} \quad (p + \rho/3) + \Lambda/3$$

After setting the cosmological constant $\Lambda = 0$ and eliminating ρ, one obtains as a solution for R (radius of universe as power law function).

$$R = (9/2GM)^{1/3} \ t^{2/3}$$

The Riemannian curvature tensor:

$$\mathbf{R}_{\mu\nu} - \tfrac{1}{2} R g_{\mu\nu} = T_{\mu\nu} + \Lambda g_{\mu\nu}$$

is often written: $\mathbf{R}^{\alpha}_{\beta\mu\nu}$

And denotes a *fourth order* tensor (of n^4 components). Lowering the contravariant indices produces a Riemann tensor of the 1st kind or:

$$\mathbf{R}_{\alpha\beta\mu\nu} = g_{\mu\gamma} \mathbf{R}^{\gamma}_{\beta\mu\nu}$$

To then briefly summarize the import of the tensor $\mathbf{R}^{\alpha}_{\beta\mu\nu}$ we note it serves as a quantitative measure of the *curvature* of spacetime. If then, $\mathbf{R}^{\alpha}_{\beta\mu\nu} \neq 0$ the spacetime is curved. Conversely, when $\mathbf{R}^{\alpha}_{\beta\mu\nu}$

= 0 the spacetime is flat. Thus, $g_{\mu\nu}$ is a metric tensor from which the Riemannian curvature tensor can be calculated[9].

Now, although $\mathbf{R}_{\alpha\beta\mu\nu}$ has 256 (e.g. 4 x 4 x 4 x4) components, only 20 are independent. At this point we may write the following useful identities:

1) $\mathbf{R}_{\alpha\beta\mu\nu} = -\mathbf{R}_{\beta\alpha\mu\nu}$ *First skew symmetry*

2) $\mathbf{R}_{\alpha\beta\mu\nu} = -\mathbf{R}_{\alpha\beta\nu\mu}$ *Second Skew symmetry*

3) $\mathbf{R}_{\alpha\beta\mu\nu} = \mathbf{R}_{\mu\nu\alpha\beta}$ *Block Symmetry*

4) $\mathbf{R}_{\alpha\beta\mu\nu} + \mathbf{R}_{\alpha\mu\nu\beta} + \mathbf{R}_{\alpha\nu\beta\mu} = 0$ *(Bianchi's Identity)*

If we consider only (1) and (2) it's clear that $R_{\alpha\beta\mu\nu}$ can be regarded as an anti-symmetric second rank tensor (in α, β) and as an antisymmetric second rank tensor (in μ, ν). Given that a second rank antisymmetric tensor has 6 independent components we are then left with no more than 6 x 6 = 36 components in the form of a 6 x 6 matrix. (The Riemann tensor is the only one that can be formed by taking linear combinations of the second derivatives of the matrix.)

---

[9] Another way to see the connections is that $\mathbf{g}_{\mu\nu}$ has a similar role to the vector potential $\mathbf{A}$ of electrodynamics. The curvature tensor plays a similar role to the $\mathbf{E}, \mathbf{B}$ fields. Thus we know $\nabla \times \mathbf{A} = \mathbf{B}$, for example.

Consider the nonzero Christoffel symbol values seen in the earlier example for the metric:

$$g_{11} = 1, \quad g_{22} = X^1, \quad g_{33} = X^2$$

Since n = 3, then only 6 components of the Riemann tensor need to be considered:

$$R_{1212}, \; R_{1313}, \; R_{1213}, \; R_{2323}, \; R_{2123} \text{ and } R_{3132}$$

Excursion: Show that the Riemann components given above are really the only ones that need to be considered. Also show that:

$$R_{1212} = g_{11} R^{\alpha}{}_{\beta\mu\nu} = g_{11} R^1{}_{212} = 1/X^1$$

And: $R_{2323} = g_{22} R^2{}_{323} = 1/X^2$

7.5. Ricci Tensor:

The Ricci tensor is completely determined by knowing the quantity $R^{\alpha}{}_{\alpha}$ for all vectors V_i of unit length. The tensor is obtained by defining a Ricci tensor of the 2nd kind thus:

$$R^{\alpha}{}_{\beta} = g^{\alpha s} R_{s\beta}$$

The number of independent components of this tensor in a space of N- dimensions is: $\frac{1}{2} N (N + 1)$.

Hence, there will be *three* components if N = 2, 6 components if N = 3 and *10 components* if N = 4. In the latter we have the case for relativistic 4 – dimensional space-time.

We consider here the simplified case for N = 2 and let the metric of interest be:

$$g_{11} = 1, \quad g_{22} = x^1,$$

For this (N=2) case:

$$R = g^{11} (R'_{11}) + g^{22} (R'_{21})$$

Where: $g^{11} = 1, \quad g^{22} = 1/x^1,$

$$R_{11} = g^{22} (R'_{21}) = (1/x^1)(-1/x^1)$$

$$R_{22} = g^{11} (R'_{12}) = (1)(-1/x^1)$$

$$R \text{ (Ricci)} = g^{11} (R_{11}) + g^{22} (R_{22})$$

$$R \text{ (Ricci)} = g^{11} (1/x^1)(-1/x^1) + g^{22} (-1/x^1)$$

$$R \text{ (Ricci)} = = (1)[-1/(x^1)^2] + (1/x^1)[-1/x^1]$$

$$= -1/(x^1)^2 - 1/(x^1)^2 = -2/(x^1)^2$$

The reader should bear in mind this is the simplest form of the calculation and while in this instance the components are proportional to the components of the metric tensor, this is not true for spaces of higher dimensionality. For example, if N= 3, one has six components and the final equation is written:

$$\mathbf{R} \text{ (Ricci)} = g^{11} (\mathbf{R}_{11}) + g^{22} (\mathbf{R}_{22}) + g^{33} (\mathbf{R}_{33})$$

Where: $\mathbf{R}_{11} = g^{22} \mathbf{R}_{2112}$

$$\mathbf{R}_{22} = g^{11} \; \mathbf{R}_{1221} + \; g^{33} \; \mathbf{R}_{3223}$$

$$\mathbf{R}_{33} = \; g^{22} \; \mathbf{R}_{2332}$$

Inquiry Exercise: Show that for computation of the Ricci tensor for $N = 3$:

$$\mathbf{R}_{ij} = \; g^{11} \; \mathbf{R}_{1ij1} + g^{22} \; \mathbf{R}_{2ij2} + g^{33} \; \mathbf{R}_{3ij3}$$

And that other nonzero components of the form

$\mathbf{R}_{aija}$ are:

$\mathbf{R}_{2112}$ and $\mathbf{R}_{3213}$ where:

$$\mathbf{R}_{3213} = -\mathbf{R}_{3132}$$

And: $\mathbf{R}_{2112} = -\mathbf{R}_{1212}$

This section would be incomplete without noting the importance of the scale factor a(t). (The metric on the spacelike slice, i.e. $\sum$ (t) – see Fig. 3.4, is given by the scale factor a(t) times the constant curvature, e.g. $\kappa = +1, 0, -1$.). Further one can then write of the "sectional curvature" $\kappa / a(t)^2$, so when:

$| \kappa | = 1$

The scale factor simply becomes the curvature radius.

Basically, a(t) is completely determined by Einstein's field equation. As shown above this is a tensor equation in space time but for the simplified isotropic and homogeneous case it reduces to an ordinary differential equation, viz.:

$$(a'/a)^2 + \kappa/a^2 = 8\pi G \rho /3$$

The first term is the Hubble parameter (not technically a constant) which indicates how fast the universe is expanding. (Currently, we estimate the corresponding constant $H_0 \approx 70$ km/ sec/Mpc). If we substitute $H = (a'/a)$ into the simplified Einstein field equation we see that when $\kappa = 0$ the mass −energy density is:

$$\rho_0 = 3H^2/8\pi G$$

Which mass density works out – using the current value of H_0, to about :

$$9.3 \times 10^{-27} \text{ kg/ m}^3$$

In effect, if we can measure the current mass density (ρ_0) and the Hubble constant we can obtain the sign of the curvature.

Inquiry Exercise 2: Show how the sign and value of the curvature κ can be obtained if one knows ρ_0 and H_0 to sufficient precision.

7.6.Parametric Representations:

In advanced treatments of curvature including determining the curvature of a complex curve, parametric representations are the norm.

Worked Problem (3): Parametric representation:

175

Given the parameters:

$x = 6 \sin 2t$, $y = 6 \cos 2t$, $z = 5t$

For which the tangent to the curve at a point is:

$\mathbf{T} = d\mathbf{R}/ds = \mathbf{i}(dx/ds) + \mathbf{j}(dy/ds) + \mathbf{k}(dz/ds)$

$(dx/dt)(dt/ds) + (dy/dt)(dt/ds) + (dz/dt)(dt/ds)$

Find an expression for $|\mathbf{T}|$ and $\mathbf{T}$ in terms of dt/ds

Solution:

$dx/dt = 12 \cos t\, 2t$, $dy/dt = -12 \sin 2t$, $dz/dt = 5$

Therefore:

$|\mathbf{T}| = 1 = (dt/ds)^2 [(12 \cos t\, 2t)^2 + (-12 \sin 2t)^2 + 5^2]$

$|\mathbf{T}| = 1 = (dt/ds)^2 [(144 + 25)] = (dt/ds)^2 (169)$

So that: $(dt/ds) = 1/13$

And:

$\mathbf{T} = 1/13\ [12 \cos t\, 2t)\,\mathbf{i} + (-12 \sin 2t)\,\mathbf{j} + 5\,\mathbf{k}\,]\, dt/ds$

The curvature can then be obtained from:

$\kappa = |d\mathbf{T}/ds|$

Thus, any curve can be given by a parametric representation: $u1 = u1(t)$ and $u2 = u2(t)$

For such a curve, consider now the distance between two infinitely near points on the surface, i.e. the distance or interval ds between (u1, u2) and (u1 + du1, u2 + du2) such as shown:

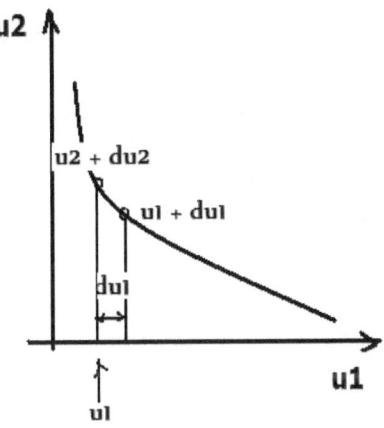

Fig. 7.5. Parametric curve

Then this is determined by: $ds^2 = dx^2 + dy^2 + dz^2$

For which we set:

$dx = \partial x / \partial u1 + \partial x / \partial u2$

corresponding to expressions for dy and dz. The quadratic differential form is then obtained:

$$ds^2 = \sum {}^3_{i,k=1} \ g_{ik} \, du_i \, du_k \quad (g_{ik} = g_{ki})$$

With coefficients:

$g_{ik} = \partial x / \partial u_i \ \partial x / \partial u_k + \partial y / \partial u_i \ \partial y / \partial u_k$

$+ \partial z/ \partial u_i \; \partial z/ \partial u_k$

The length of any arbitrary curve given by: $u_1 = u_1(t)$ and $u_2 = u_2(t)$ is:

$$\int ds = \int [\Sigma_{ik} g_{ik} du_i/ dt \; du_k/ dt]^{1/2} \; dt$$

This may be compared with the form:

$$\mathbf{ds} := \int_{x1}^{x2} \sqrt{1 + \left(\frac{d}{dx}f(x)\right)^2} \; dx$$

For the usual arc length obtained in x, f(x).

7.7 Other Observational Aspects of General Relativity

1. Slowing Clocks in Gravitational Fields.

Einstein's general theory of relativity predicts that clocks *run slower in gravitational fields* (a phenomenon called 'gravitational time dilation') In this case, for the Earth, one would have the fractional difference in proper time, as a fraction of time passage t[10]:

$$\Delta\tau/ \tau \approx \; d\tau_2 - d\tau_1/ dt \; \approx \; GM(1/r_1 - 1/r_2)$$

---

[10] Cf. Ohanian, H.C. and Ruffini, R. (1995), :**Gravitation and Spacetime**, p. 182.

where G is the Newtonian gravitational constant, M is the Earth's mass, and g is the acceleration of gravity (g = 980 cm/ sec² in cgs) and c = 3 x 10¹⁰ cm/sec.

In terms of frequency (or clock tick rate) this amounts to a shift in frequency. A large elapsed time ($d\tau_2 - d\tau_1$) implies a fast tick rate so high frequency. This shows the difference in frequency must be proportional to the elapsed time so:

$$\Delta v/ v = (v_2 - v_1)/ v = \Delta\tau/ \tau \approx GM [1/r_1 - 1/r_2]$$

Generalizing, any atom placed inside a significant gravitational potential will display a redshift compared with a similar atom outside the potential. Thus, inside the potential (at smaller radius r_1) the tick rate slows compared to that measured at radius r_2. This means the frequency is slower or redshifted.

Near to Earth's surface it is common to make the approximation:

$$GM [1/r_1 - 1/r_2] \approx GM \Delta r/ R^2 = g \Delta r$$

Let us say that in one experiment ($r_2 - r_1$) = 0.001 mm = 0.0001cm, then:

$$\Delta\tau/ \tau \sim (980 \text{ cm/s}^2)(10^{-4} \text{ cm})/ (3 \times 10^{10} \text{ cm/sec })^2$$

$$\Delta\tau/ \tau \approx 10^{-22}$$

and for an interval say t = 0.01 sec, $\Delta\tau$ =

$$(10^{-22})(0.01 \text{ sec}) = 10^{-24} \text{ sec}$$

7.8. Deflection of Light in Gravitational Field

The deflection of star light in a gravitational field was tested during solar eclipse in 1919, and was actually described in Eddington's book in detail[11]. A rough illustration of the effect is shown below:

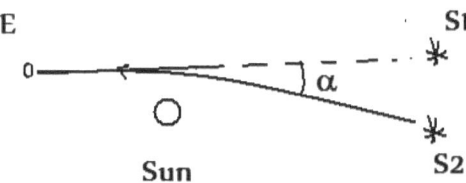

Fig.7.6. Deflection of star light

In the diagram the light from the star at actual position S2 is seen to deflect by some angle α, thereby altering the image position to that seen at S1. This is a direct result of the effect of the gravitational field of the Sun on the light rays. The true direction is thus alone the ray ES2 while the deflected position is along the ray ES1.

Theoretically and quantitatively, one get obtain an estimate of the magnitude of deflection by incorporating another parameter – call it **b** – as shown in the diagram below:

---

[11] Eddington, A.S., op.. cit., p. 102.

180

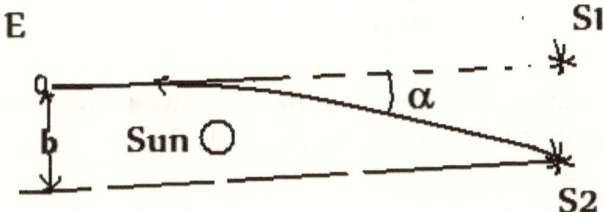

Fig. 7.7. Determining the linear deflection, b.

Then we obtain for the deflection angle, α:

$\alpha = -\,4\,GM/\,b$ or (in cgs units):

$\quad \alpha \;\; = -\,4\,GM/\,b\,c^2$

Einstein, in his own paper: 'On the Influence of Gravitation on the Propagation of Light', gives the result as: $\alpha = 2k\,M/\,c^2\,\Delta$

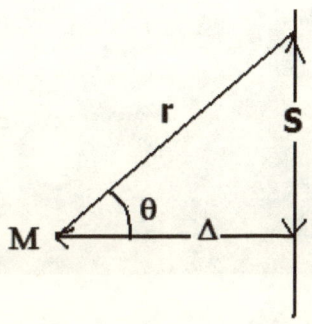

Fig. 7.8. Diagram for Einstein's derivation of α.

Here k = G, the gravitational constant and Δ = the distance of the ray from the center of the body. It is obtained from the diagram of Fig. 12.

Einstein used the cosine of the angle and integrated from the negative angle θ = - π/ 2 to θ = π/ 2:

$$\alpha \;=\; 1/\,c^2 \int_{-\pi/2}^{\pi/2} k\,M/\,r^2 \; \cos\theta \; ds$$

7.9. Advance of Mercury's Perihelion

Another prediction from General Relativity is the advance of Mercury's perihelion. This is a phenomenon in no way predictable from Newtonian gravitation and is illustrated below, with the orbits separated in exaggerated fashion:

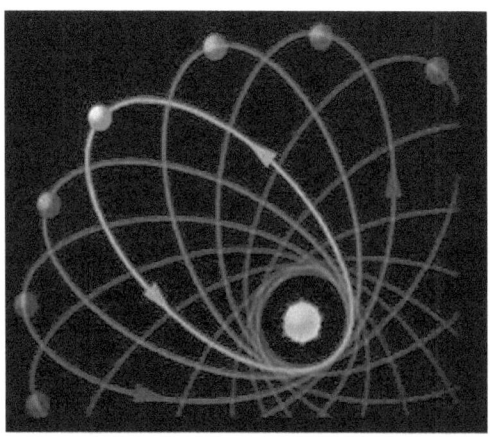

Fig. 7.9. Advance of Mercury's Perihelion

The amount of rotation of 'the planetary ellipse' due to the effects of gravitation in general relativity was set out in the following equation of Einstein's:

$$\varepsilon = [24\,(\pi)^3\,a^2]/\,T^2\,c^2\,[1 - e^2]$$

where ε is the advance (or rotation) in seconds of arc, T is the period of revolution in seconds, c the velocity of light and e the eccentricity of the orbit.

Einstein, on page 164 of the same paper, asserts that for Mercury $\varepsilon = 43$ seconds per century

Worked Problem:

Validate the magnitude of the perihelion advance of Mercury from Einstein's given equation.

Solution: $c = 3 \times 10^{10}$ cm/s

Einstein framed his equation in cgs units, so we must have:

$a = (1.5 \times 10^{13}$ cm$)\,(0.387) = 5.8 \times 10^{12}$ cm

Based on using 1 astronomical unit (AU) = 1.5×10^{13} cm. But from a Table of distances, Mercury's semi-major axis a = 0.387 AU.

The period, T (in seconds) is just the length of Mercury's year (in days = 87.96, again from a Table) multiplied by the seconds-length of an Earth day, or 86,400 s:

$T = 7.6 \times 10^6$ s

The eccentricity, e from a similar Table is e = 0.205.

Substituting all these values into the given equation yields:

$\varepsilon = 5.036 \times 10^{-7}$ radian

To get the equivalent seconds of arc (or arcsec) we use 1 rad (radian) = 57.3 degree where one degree has 3600 seconds. Thus, 1 radian will have:

2.063×10^5 arcsec

So, the associated arcsec for ε will be:

$(5.036 \times 10^{-7}$ rad$) \times (2.063 \times 10^5$ arcsec/ rad$) =$

0.104 arcsec

This is the quantity defined per **century**.

At this point, you need to recall the PERIOD of Mercury is 0.2405 YRS.

The number of arcsec of perihelion advance per Earth years is:

0.104 arcsec/ 0.2405 years = 0.432

So: 43 seconds of arc per century

End of Chapter Problems:

1) Provide a matrix which satisfies:

$$\mathbf{i}^\wedge \, \mathbf{i}^\wedge + \mathbf{j}^\wedge \, \mathbf{j}^\wedge + \mathbf{k}^\wedge \, \mathbf{k}^\wedge = 7/2$$

Verify that: $\Gamma_{ijk} = \Gamma_{jik}$

2) Write out the trace for the metric tensor.

$\mathbf{g}_{ik} =$
$(1.....0..............0)$

$(0.....r^2..............0)$

$(0.....0......r^2 \sin\theta \,)$

Is $\mathrm{Tr}(\mathbf{g}_{ik})$ the same for the inverted matrix?

3) Let: $a_{ij} := \begin{pmatrix} 2 & 0 & 3 \\ 5 & 1 & 2 \\ 4 & 5 & 7 \end{pmatrix}$

And $\mathbf{x}_i$ and $\mathbf{y}_j$ be first order tensors (vectors) respectively given by:

$\mathbf{x}_i = (2, 1, 4)$

$\mathbf{y}_j = (3, 7, -1)$

a)Find: $\mathbf{a}_{ij}\, \mathbf{x}_i \;+\; \mathbf{a}_{ji}$

b)Find: $(\,\mathbf{a}_{ij}\; - \; 2/3 \;\delta_{ij}\,)$

4) Given a 2nd order tensor $\mathbf{a}_{ij}$ defined by the matrix:

$$a_{ij} := \begin{pmatrix} 2 & 0 & 3 \\ 5 & 1 & 2 \\ 4 & 5 & 7 \end{pmatrix}$$

$$x_j := \begin{pmatrix} 2 \\ 1 \\ 4 \end{pmatrix}$$

And:

Find: $\mathbf{a}_{ij} \cdot \mathbf{x}_j$

4) For the same 2nd order tensor $\mathbf{a}_{ij}$ and:

$$\delta_{ij} := \begin{pmatrix} 1 & 0 & 0 \\ 0 & 1 & 0 \\ 0 & 0 & 1 \end{pmatrix}$$

Find: $\mathbf{a}_{ij} \cdot \delta_{ij}$

Obtain the **trace** of the matrix $\mathbf{a}_{ij} \cdot \delta_{ij}$

5) Decompose the matrix:

$$a_{ij} := \begin{pmatrix} 2 & 3 & 2 \\ 5 & 7 & -2 \\ 4 & -4 & 0 \end{pmatrix}$$

into its *symmetric* and *anti-symmetric* parts

6) Write out in long form the full sum (all terms) for the interval $ds^2 = g_{\mu\nu} dx^\mu dx^\nu$

Be sure to include *all terms* for the sum that are applicable to standard form. (Hint: Bear in mind for the components of the metric with *repeat subscripts*, e.g. g_{33} the coefficient for the term is unity, and for all other terms the coefficient is 2.)

7) Show that for the Ricci tensor, for $N = 3$:

$$R_{ij} = g^{11} R_{1ij1} + g^{22} R_{2ij2} + g^{33} R_{3ij3}$$

8)(a)Assume a negative Gaussian curvature. Write out the components of the metric $g_{\mu\nu}$ and the full sum (all terms) for the relevant interval:

$$ds^2 = g_{\mu\nu} dx^\mu dx^\nu$$

Repeat this for a *positive* (+1) Gaussian curvature.

9) Using the metric: $G =$

(12....4....... 0)
(4....1.......... 1)
(0....1... $(x^1)^2$)

Calculate the length of the curve given by:

$x^1 = 3 - t$, $x^2 = 6t + 3$, $x^3 = \ln t$

Where: $1 \le t \le e$

10) For the parametric example problem with:

$\mathbf{T} = 1/13$ [12 cost 2t) $\mathbf{i}$ + (-12 sin 2t) $\mathbf{j}$ + 5 $\mathbf{k}$] dt/ds,

Find the curvature and the *length of the curve* from:
$t = 0$ to $t = \pi$.

(11) Consider a curve for which:

$r(\theta) = \theta - \sin(\theta)$

a) Find the arc length between $\theta = -\pi$ and $\theta = \pi$.

b)Find the mean curvature, κ, at three arbitrary points on the curve.

12) A curve is given in spherical coordinates x^i by:

$x^1 = t$, $x^2 = \arcsin 1/t$, $x^3 = (t2 - 1)^{1/2}$

Compute the length of the arc between $t = 1$ and $t = 2$

13) Under the metric:

G =

[1   cos 2x^2]

[cos 2x^2 1]

a) Compute the **norm** of the vector $\mathbf{V} = dx^i/dt$ evaluated along the curve: $x^1 = -\sin 2t$, $x^2 = t$.

b) Find the arc length between $t = 0$ and $t = \pi/2$

14) Obtain the *Ricci* tensor for the metric :

$$g_{11} = 1, \quad g_{22} = x^1, \quad g_{33} = x^2.$$

15) Let $\boldsymbol{\Phi}$ be some generic tensor. Prove that Stokes theorem applies as follows:

$$\int \boldsymbol{\Phi} \cdot dS = \int \int \nabla \times \boldsymbol{\Phi} \cdot dS$$

16) Find the *normalized* eigenvectors of the 3 x 3 determinant:

(1...0....6)
(0..-2...0)
(6....0..6)

17) Calculate the current value of the curvature of the universe using: the current Hubble constant H_o and the density: $\rho_o = 9.3 \times 10^{-27}$ kg/ m^3

18) Consider the *nonzero* Christoffel symbol values for the metric:

$$g_{11} = 1, \quad g_{22} = 2x^1, \quad g_{33} = 3x^2$$

Thence or otherwise compute:

$$\Gamma^1{}_{22} , \quad \Gamma^2{}_{12} , \quad \Gamma^2{}_{21}$$

$\Gamma^3{}_{23}$, $\Gamma^3{}_{32}$

And explain why one or more values may be the same.

19) From the values of the metric in (19) compute the Riemann tensor components:

a)$g_{11} R^1{}_{212}$ b) $g_{22} R^2{}_{323}$

c) Show: $R_{\alpha\beta\mu\nu} + R_{\alpha\mu\nu\beta} + R_{\alpha\nu\beta\mu} = 0$

20) Write out in long form the full sum (all terms) for the interval $ds^2 = g_{\mu\nu} dx^\mu dx^\nu$

(21) (a) Assume a negative Gaussian curvature. Write out the components of the metric $g_{\mu\nu}$ and the full sum (all terms) for the relevant interval $ds^2 = g_{\mu\nu} dx^\mu dx^\nu$

(b) Repeat this for a positive (+1) curvature.

22) a)Using the appropriate relations, estimate the density of the universe at a time 0.01 second after the Big Bang

b) Repeat your computation if the Hubble constant is found to be H = 100 km/ sec/Mpc.

23) Using Einstein's relation for the deflection of starlight in a strong gravitational field:

$\alpha = 1/c^2 \int_{-\pi/2}^{\pi/2} k M/r^2 \cos \theta \, ds$

Show that this would be equal to: $\alpha = 2 \Phi/c^2$

VIII. *EPR Paradox & Quantum Nonlocality*

8.1. The EPR Paradox

In 1935, Einstein along with two colleagues, Boris Podolsky and Nathan Rosen, devised a thought experiment.[12] This has since been called the EPR experiment based on the first initials of their names. Einstein, Podolsky and Rosen (E-P-R) imagined a quantum system (atom) which could be ruptured such that two electrons were dispatched to two differing measurement devices. Each electron would carry a property called 'spin'. Since the atom had zero spin, this meant one would have spin (+ 1/2), the other (- 1/2). The diagram below illustrates this, the atom being disrupted inside the box, with its opposing spin electrons sent to the left and right.

(+1/2) ↑ <---------|------------> ↓ (-1/2)
             [BOX]

Orthodox quantum mechanics forbade the simultaneous measurement of a property (say different spin states) for the same system. If you got one, you could not obtain the other. This was a direct outcome of the Heisenberg Indeterminacy Principle which stated that simultaneous quantum measurements could not be made to the same precision.

E-P-R argued that this showed the incompleteness of quantum mechanics. It was not the 'paragon' of physical theories its apologists claimed, if such

---

[12] Einstein, Podolsky, and Rosen.: *Physical Review*, 777.

indeterminacy was fundamentally embedded within it. At the same time they conceived an 'experiment' in which both spins could be identified - with the sole assumption that both were in definite states from the instant of their parent atom's disruption.

Then, we need only know one spin to obtain the other. Say we know or can measure one spin to be $(+1/2)$.[13] Since the total atomic spin is zero, the other electron must have spin $(-1/2)$ since: $(-1/2) + (+1/2) = 0$

Thus, we manage to skirt the Indeterminacy Principle, and obtain both spins simultaneously without one measurement disturbing the other. We gain completeness, but at a staggering cost. Because this simultaneous knowledge of the spins implied that information would have had to propagate from one spin measuring device (on the left side) to the other (on the right side) instantaneously! This was interpreted to mean faster-than-light communication, which violates special relativity. In effect, a 'paradox' ensues: quantum theory attains completeness only at the expense of another fundamental physical theory - relativity.

By this point, Einstein believed he finally had Bohr by the throat. Figuring Bohr might come up with some trick or sly explanation, Einstein went one better at the 6th Solvay Conference held in 1930, actually designing a device (see image on next page) that he was convinced would have Bohr in tears trying

---

[13] More technically, this is what is referred to as 'the z-component of electron spin', since the electron is visualized as a spinning top, with z-axis (i.e. component) in the axial or z-direction.

to find a solution. (According to reports, it very nearly did, and a number of participants insisted Neils was in a state of "shock" believing there was no real solution.)

Using his hypothetical (purely on paper) thought device, Einstein wanted to put to rest once and for all the notion that quantum mechanics was complete, or was in any way a proper science. The mechanical device contrived by Einstein was designed as a counter-example to the Heisenberg Uncertainty principle for energy and time which states:

$$\Delta E \, \Delta t \geq h/2\pi$$

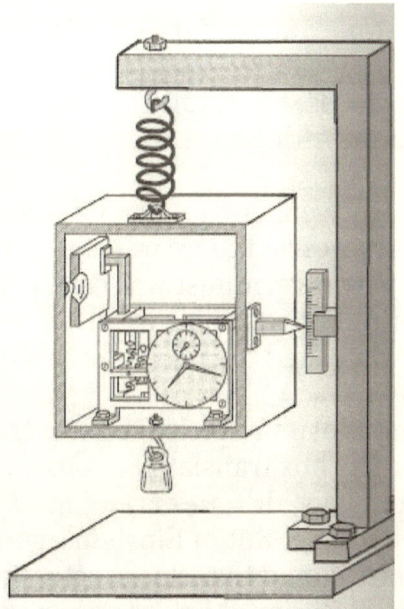

Fig. 8.1: Einstein's Thought Experiment Device

In the device, a weight scale is located and one can

see it when a door (front of box) opens, with the door controlled by a clock timer. Whenever the door flaps open, even for a split second, one photon escapes and the weight difference (between original box and after) can be computed using Einstein's mass-energy equation, e.g.: m = E/ c². Thus, the difference is taken as follows:

Weight(**before door opens**) - weight (**after** - with 1 photon of mass m = E/ c² gone)

Thus, since the time for brief opening is known (*Δ t*) and the photon's mass can be deduced from the above weight difference, Einstein argued that one can in principle find both the photon's energy and time of passage to any level of accuracy *without any need for the energy-time uncertainty principle.*

In other words, the result could be found on a totally deterministic basis!

Bohr nearly went crazy when he studied the device, and for hours worried there was no solution and maybe the wily determinist was correct after all. When Bohr did finally come upon the solution, he realized he'd hoisted the master on his own petard.

The thing Einstein overlooked was that his very act of weighing the box translated to observing its displacement (say, dr = r2 - r1) *within the Earth's gravitational field*. But in Einstein's general theory of relativity, he'd found that clocks actually *do run slower in gravitational fields* (a phenomenon called 'gravitational time dilation') In this case, for the Earth, one would have the fractional difference in proper time, as a fraction of time passage t:

$$dt/t \approx GM(1/r_1 - 1/r_2) \approx g(dr)/c^2$$

where G is the Newtonian gravitational constant, M is the Earth's mass, and g is the acceleration of gravity (g = 980 cm/ sec^2 in cgs) and c = 3 x 10^{10} cm/sec.

Let us say the box deflection (r2 - r1)was 0.001 mm = 0.0001cm, then:

$$dt/t \sim (980 \text{ cm/s}^2)(10^{-4} \text{ cm})/ (3 \times 10^{10} \text{ cm/sec })^2$$

$$dt/t \approx 10^{-22}$$

and for an interval say t = 0.01 sec, dt =

$$(10^{-22})(0.01 \text{ sec}) = 10^{-24} \text{ sec}$$

In other words, the observation would actually generate a time uncertainty of 10^{-24} sec- and hence an uncertainty dE in the energy of the photon. In other words, after the displacement (r2 - r1) arising from the measurement, the clock is in a gravitational field *different* from the original one. (The *Energy uncertainty* can meanwhile be computed from the Heisenberg energy -time relation to be dE $\approx 10^{-10}$ J)

Quantum theory prevails again!

Einstein's challenges to Bohr in the aftermath were all kind of half-hearted and had nowhere near the intensity of his clock-door device work of art. Rather than join happily with other QM theorists at the last Solvay Conference in 1933 Einstein - the perpetual determinist- remained on the sidelines "feeling the same uneasiness as he had before".

He went to work separately, on a "unified field theory" while the quantum theory edifice was formulated to its present maturity without him.

Aside: Most people outside physics are unaware there were seven Solvay Conferences in all, in the course of which the essential underpinnings and core interpretation of quantum mechanics was thrashed out – leading to the *Copenhagen Interpretation*.

In the orthodox Copenhagen (and most conservative) interpretation of quantum theory, there can be no separation of observed (e.g. spin) state until an observation or measurement is made. Until that instant (of detection) the states are in a superposition, as described above.

More importantly, the fact of superposition imposes on all quantum phenomena an inescapable 'black box'. In other words, no information other than statistical can be extracted before observation.

8.2. Bell's Theorem and the Aspect Experiment

Years later, mathematician John S. Bell asked the question: 'What if the E-P-R experiment could actually be carried out? What sort of mathematical results would be achieved?' In a work referred to as *"the most profound discovery in the history of science"*, Bell then proceeded to place the E-P-R experiment in a rigorous and quantifiable context, which could be checked by actual measurements.

In a landmark theoretical achievement in 1964, Bell formulated a ***thought experiment*** based on a design similar to that shown at the opening of the chapter. He made the basic assumption of locality (i.e. that no communication could occur between A1 and A2 at any rate faster than light speed). In what is now widely recognized as a seminal work of mathematical physics, he set out to show that a theory which upheld locality could reproduce the predictions of quantum mechanics. His result predicted that the above sum, S, had to be less than or equal to 2 ($S \leq 2$). This is known as *the Bell Inequality*.

To test quantum conformity to Bell's Theorem, Alain Aspect and his colleagues at the University of Paris, set up an arrangement as sketched below.[14] In these experiments, the detection of *the polarizations*[15] of photons was the key. These were observed with the photons emanating from a Krypton-Dye laser and arriving at two different analyzers, e.g.

```
P1 ↓| <------------|-------------> |↑ P2
   A1           D            A2
```

Here, the laser device is D, the analyzers (polarization detectors) are A1 and A2 and two representative polarizations are given at each, for two photons P1 and P2. The results of these remarkable experiments disclosed apparent and instantaneous connections between the photon polarizations at A1 and A2. In the case shown, a photon (P1) in the

---

[14] Aspect, Grangier, and Roger: *Physical Review Letters*, 91.

[15] Polarization is the orientation in space of the electric field $\underline{E}$, associated with light. This can be altered, subject to the imposition of different filters and devices.

minimum (o) intensity polarization mode, is anti-correlated with one in the maximum intensity (1) mode.

Say, twenty successive detections are made and we obtain, at the respective analyzers (where a '1' denotes spin +1/2 detection and 'o' spin (-1/ 2):

A1:   1 0 1 0 1 0 1 0 1 0 1 0 1 0 1 0 1 0 1 0
A2:   0 1 0 1 0 1 0 1 0 1 0 1 0 1 0 1 0 1 0 1

On inspection, there is a 100% anti-correlation (i.e. 100% negative correlation) between the two and an apparent nonlocal connection. In practice, the experiment was set out so that four (not two - as shown) different orientation 'sets' were obtained for the analyzers. These might be denoted: (A1,A2)I, (A1,A2)II, (A1,A2,)III, and (A1,A2)IV.

Each result is expressed as a mathematical (statistical) quantity known as a 'correlation coefficient'.[16] The results from each orientation were then added to yield a sum S:

S = (A1,A2)I + (A1,A2)II + (A1,A2,)III + (A1,A2)IV

In his (1982) experiments, Aspect determined the sum with its attendant indeterminacy to be:[17] S = 2.70 ± 0.05 and in so doing experimentally validated Bell's Inequality.

---

[16] For example, if a set of data: 1, 1, 1, 1 is correlated with another set: 0.5, 0.5, 0.5, 0.5, the correlation coefficient is 1.0. The range is between 0 (no correlation) and 1.0 (perfect correlation).

[17] Aspect, A. et al, *op. cit.*

The crucial significance of the Aspect experiment (1982) and earlier the Clauser experiment in 1969, had not been lost on physicist David Bohm. Both experiments reinforced for him that a novel concept had to be introduced to account for these nonlocal results. The old answer of the Copenhagen theorists: *'Leave it alone!'*, wouldn't do any more. In essence, Bohm had become convinced that the entrenched vagueness and philosophical abdication of the standard Copenhagen Interpretation could no longer be supported.

8.3. Attempts to Refute the Experimental Basis of Quantum Nonlocality:

It wasn't long before a number of attacks and objections were made by (Copenhagen interpretation favoring) physicists determined to retain locality. The late physicist Heinz Pagels, in a 1982 book[18], attacked the basis for *real nonlocality* by arguing that the separate polarization records at each analyzer are themselves totally random sequences.[19] Hence, one cannot obtain any useful information except by correlating two sets of records in the manner shown.

However, at genuinely vast distances - say 10 light years between A1 and A2 - such correlation can never be practicable. In that sad event we are left with either one totally random record (say for A1) or another, but with absolutely no prospect of comparing the two and getting positive information. In this way, Pagels insists, real nonlocality is avoided.

---

[18] Pagels, *The Cosmic Code.*
[19] Pagels,: *op. cit.*, 144- 152.

But is it? One can object first to Pagels' exclusion of a real nonlocal influence on the basis of rejecting definite polarizations before the measurement.[20] This excludes all null measurement techniques. For instance, say we know a polarizer will yield two processes, one of reflection, r> and one of transmission, t>.

Then: $P(U>) = r> + t>$

If then only $r>$ is observed at some detector after interaction with P, it is concluded $t> = 0$. But this is by inference, not actual measurement.

While this is consistent with the Copenhagen Interpretation of Quantum Mechanics it is not consistent with any other interpretation. For example, from the standpoint of the *Stochastic Interpretation of Quantum Mechanics* (SIQM), Pagels' exclusion is inapplicable. A mathematical treatment of the Aspect results from an SIQM perspective shows that the photon polarizations can be considered as "*always* to have been defined and nonlocally connected".[21]

One can also attack Pagels' objection based on his definition of randomness. He uses the same definition as Richard von Mises, a German mathematician.[22] In von Mises' conception, randomness emerges as an *irreducible lawlessness*.[23] That is, it is defined without regard for regularity of place or pattern. Let

---

[20] Pagels,: *op. cit.*, 150.

[21] Cufaro-Petroni and Vigier *Physics Letters*, (93A), 383.

[22] Pagels,: *op. cit.*, 91.

[23] von Mises: *Probability, Statistics and Truth*, 23-25.

me give an example, in which a coin is tossed ten times in a row and heads (H) and tails (T) registered:

T H T H T H T H T H

Is the sequence random? Based on only ten tosses it is difficult to say. However, a pattern of sorts appears to have emerged: each H and T alternate over one cycle. In von Mises' definition this isn't random, since a given event (in this case T or H) is required to be independent of place. This means I'm unable to predict the next place for an H or T (which is to say, the next event.) Gaming tables in Las Vegas are based on this principle.

The key point, missed by Pagels is that even the assertion of what constitutes randomness is based on an idea of what does not. In other words, a non-random pattern or regularity is required to define randomness! There must be an awareness of what is *non-random* before there can be awareness of what is random!

Since a unique definition of randomness is nonsense, or at least meaningless, one cannot maintain a priori (as Pagels does) that a sequence of polarizations will necessarily be a random sequence. The reason is that this assertion demands vastly more information than can be accessed in the experiment. Even a seemingly endless random sequence could merely be the subset of a vastly larger orderly sequence. So - something like: 0110111111001

That appears apparently random, but could be a **subset** of the orderly sequence:

0110111111001 **0110111111001** 0110111111001 ....
(One hundred times)

It is impossible to know the pattern *unless the totality of information is available*. Unfortunately, the observer is never in a god-like position to state categorically one way or the other, by examining one record. However, a priori there is no reason why the nonseparability disclosed in the Aspect experiment should not apply over a distance of ten light years or more. The quantum potential, in fact, permits this.

The erroneous assumption that nonlocality equates to faster-than-light propagation as in a signal concept. It does not. Nor do Aspect's experiments (or any others) show a faster-than-light (FTL) connection! What they show is not superluminal *transfer* of information, but rather *pre-existing connections in a higher dimensionality*! This is totally different, since it doesn't require *separate localizations* from which FTL signals emanate.

The point is that the two photons detected by A1, A2 in the Aspect experiment are *already connected in a higher dimensionality*, not readily accessible to us. The experimental results unequivocally show this, but we insist on using fragmentary language to refer to two photons - one at each analyzer, as if they are distinct entities separated by distance.

What critics perceive as some kind of relativistic cheating is really the normal behavior of a *higher dimensionally- unified entity.*[24]   This is why the notion of superluminal signals, or indeed any signals, serves no useful purpose in discussions of holism vs. locality. In fact, it only serves to confuse the issues.

According to Einstein's theory, nothing travels faster than light.   For example, consider the mass relation in special relativity:

$$m = m_0 / [(1 - v^2/c^2)^{1/2}]$$

Where m is the mass moving at velocity v, m_0 is the rest mass, and c the speed of light. If the mass were to move at the speed of light, c:

$$m = m_0 / [(1 - c^2/c^2)^{1/2}] =$$

$$m_0 / [(1 - 1)^{1/2}] = m_0 / 0 = \infty$$

In other words, the mass would have to literally be *infinite*, i.e. have infinite inertia, in which case it would be impossible to move. If this applies at the velocity of light threshold, it clearly rules out faster than light masses too, hence superluminal transfer of information. But this is *not* what quantum nonlocality means! In Aspect's experiment it means[25] *causally —bound photons connected by causal action*

---

[24] In any event, Stenger appears not to be aware that superluminal travel is not expressly forbidden by general relativity. It is only precluded by the gravity-free flat space-time of special relativity. For more on this, see: Parsons: Science, (274), 202.

[25] Aspect, Grangier, and Roger,: *Physical Review Letters* (47), 460.

at a distance which is different.

As Alain Aspect et al put it[26];

No energy can be exchanged between the photons in Σ_0, so that no causal anomaly results from this particular action at a distance.

This is known as *causal covariant action at a distance* which implies simultaneous action of the observables owing to their already existing physical connection. This can be represented in the following way:

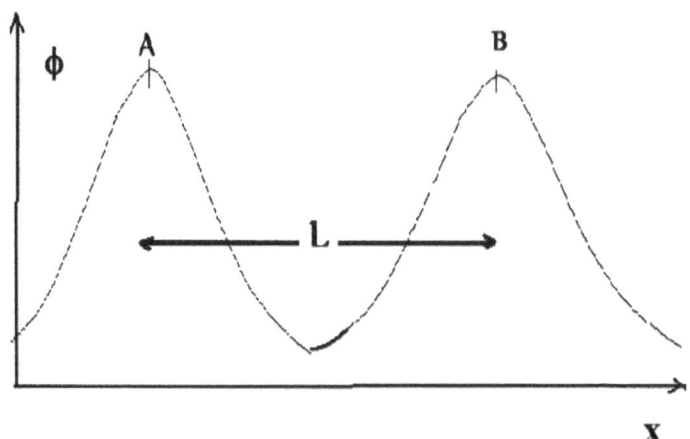

Fig. 8.2: *Apparently distant photons connected at lower levels of field intensity*

Here, two photons A and B are apparently displaced from each other by some distance, L,

[26] Aspect et al, *ibid.*

measured between their maxima, are actually connected at low levels of field intensity, ϕ. It is only at high levels of this intensity we observe them as separate (singlet) photons. It is in that displaced condition nonlocality critics usually consider superluminal transfers but in reality it's not needed because the two photons *are already bound*. Here, physicist Bernard d'Espagnat's words are certainly worth a look[27]:

The experimental corroboration of nonseparability (i.e. nonlocality) quite obviously constitutes a strong argument against the hypothesis of objective realism as applied to microscopic objects, and even....against that of objectivist realism applied to macroscopic objects only.

d'Espagnat further concedes that for anyone *"believing in both locality and realism"* - the observed violation of the Bell inequalities *"would remain inexplicable"*[28].

The preceding examples enable us to recognize that distinct levels exist for which separate physical laws have their validity of operation. Having relativity valid and operative at one, autonomous localized level in no way vitiates higher-dimensional laws at a more implicate, nonlocal level. As physicist Sunny Auyang remarks:[29]

In physics, two isolated systems are idealizations. The world is an interactive system. We ideally carve out discrete entities by neglecting weak couplings. In

---

[27] d'Espagnat.:*op. cit.*,143.
[28] *Ibid.*
[29] Auyang,: *How Is Quantum Field Theory Possible?*, 120.

most cases, the neglected coupling is weak enough so that the approximation of discrete entities is good.

I would also like to suggest that special relativity (Chapters 1-3) , which prohibits faster-than-light transfer, is a theory that ignores the weaker couplings of higher dimensionality that give rise to holism. In much the same way, Newtonian gravitation is perfectly valid and works well when the weaker couplings of general relativity are ignored, say by confining attention purely to satellites in Earth orbit.

Inevitably, problems arise when there is a conflation between two distinct regimes/couplings, and the relevant factors in each are ignored or omitted. This is bound to continue so long as the hyper-dimensional aspects of nonlocal quantum experiments remain unrecognized.

Again, the claimed choice of "reality over nonlocality" is largely a contrived problem. Bell's Theorem, like the premise of 'interfering postulates', merely shows that two or more fundamental requirements cannot co-exist in the same quantum description (Bohr's Complementarity Principle). This applies, for example, to local and nonlocal descriptions, or modeling. While I can certainly invoke one or the other, I can't use both in the same model.

One last remarkable side note: Evidently physicists associated with Canada's National Research Council in Ottawa have succeeded in actually measuring a quantum wave function (hitherto believed impossible since it was only said to be statistical in nature.). This was reported in the June 9, 2011 issue of *Nature*.

Something to consider when hard core Copenhagenites insist *"the wave function isn't real"*.

8.4. Bohm's Version of Quantum Mechanics

Regarding the various violations of the Bell Inequality, David Bohm with colleague Brian Hiley developed an alternative form of quantum mechanics to integrate it within existing observations. This began by treating the de Broglie wave as a physically real entity not merely a statistical one. Also, Bohm and Hiley refined the concept of the pilot wave, originally proposed by Louis de Broglie

If matter waves, or de Broglie waves, can be "piloted" then clearly they will have far more theoretical impact than if mere products of random encounters or observations. Hence, Bohm and Hiley 's development of a "pilot wave" theory to accompany the acceptance of B-waves physical reality.

To understand the "clocking" guidance system for these waves, we begin with the basic energy definition for the quantum given some rest frequency, f_o.

Then the energy quantum associated with this frequency is, by Planck's equation:

$E_o = h f_o$

Where h is the Planck constant. Then we can also write:

$f_0 = E_0 / h$

In the relativistic limit, for photons: $f_0 = m_0 c^2 / h$

Now, change to angular frequency ω_0 to make the synchronous mechanism consistent with that proposed by de Broglie and Bohm, Hiley[30]. Then:

$2\pi f_0 = m_0 c^2 / h$

Replacing the Planck constant by $\hbar = h / 2\pi$, the Planck constant of action:

$2\pi f_0 = m_0 c^2 / \hbar$

Then:

$\omega_0 = m_0 c^2 / \hbar$

Which is the "clock frequency" in the photon rest frame. There is also an additional condition, known as the Bohr-Sommerfield condition, for the clock to remain in phase with the pilot wave:

$$\oint p \, dx = n \hbar$$

Now, the momentum $p = m_0 c$, so that the integral becomes:

$2\pi x (m_0 c) = n \hbar$

---

[30] Bohm and Hiley: *Foundations of Physics*, (12), No. 10, p. 1001.

And the de Broglie wavelength emergence ($\lambda_D = h/p$) is evident in the equation. In this sense, we have:

$$2\pi x = n(h/m_0 c) = n \lambda_D$$

Or the same expression ($2\pi r = n \lambda_D$) we saw earlier (Chapter 4) for the standing waves in an atom. In Bohm's own development, the procession of B-waves is actually enfolded within a "packet" of P-waves[31]. A basic diagram of the arrangement is shown below.

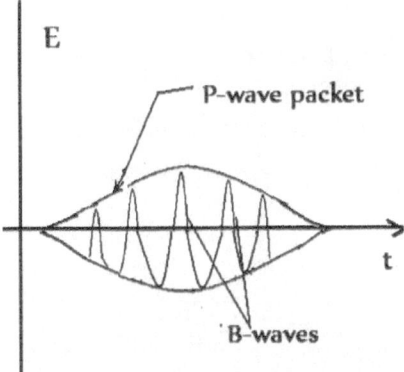

Fig.8. 3: B-waves enfolded in P-wave packet

The axis labeled E is actually the real part of the electric field component, E_z. The width of the p-wave packet is denoted by the spread:

$$\Delta k = \pi / (x - x_0)$$

Where x_0 denotes the center point of the wave packet. In other words, if the center point is at $x_0 = 0$, the packet width is just $\Delta k = \pi / x$.

[31] See, e.g. Bohm, D: *Quantum Theory*, Dover, p. 169, 1951.

The wavelength, $\lambda = 2\pi / \Delta k$ then is *much less* than the width of the packet. E.g. if $\Delta k = \pi / x$ then $\lambda = 2\pi / (\pi / x) = 2x$. so if $x = 1$ nm, then $\lambda = 2$ nm and:

$$\Delta k = \pi / x = \pi / (1 \text{ nm}) = \pi \text{ nm, but } \pi \text{ (nm)} > 2 \text{ nm.}$$

The maximum of the wave packet is approximated closely by the square of the amplitude:

$$[\, E_z \,]^2 = \quad 4 \sin^2 \Delta k \, (x - x_0) / (x - x_0)^2$$

We can check the limits of the preceding. Let $x_0 = 0$ then:

$$[\, E_z \,]^2 = \quad 4 \sin^2 \Delta k \, x / x^2$$

And:

$$[\, E_z \,]^2 = \quad 4 \sin^2 \Delta k / x$$

Conversely, let $x_0 = x$, then:

$$[\, E_z \,]^2 = 0 = 4 \sin^2 \Delta k \, (x - x) / (x - x)^2$$

Thus, the p-wave packet ceases to exist as a discrete or localized entity and thereby **loses its particle properties**. But what about the photon mass?

Mass can be derived as a basis for the wave packet spread $\Delta x = (x - x_0)$. Thus, a "particle" is represented as a finite wave packet with wavelength-based spread $\Delta\lambda$, such that, according to the Heisenberg Uncertainty Principle:

$$\Delta x = \Delta k = \Delta \left(1/\lambda \right) \sim 1$$

Now: $\lambda = \hbar / m\, v\gamma$

Where: $\gamma = \left(1 - v^2/c^2 \right)^{-\frac{1}{2}}$

Recall we saw the basic clock frequency (in the zero reference frame):

$$\omega_0 = m_0\, c^2 / \hbar$$

Which may now be generalized for relativistic speeds:

$$\omega = m\, \gamma\, c^2 / \hbar$$

In terms of the Compton wavelength:

$$= \hbar / mc$$

$$\omega = \gamma \left(c/\xi \right)$$

Given a spread in the velocity $\Delta v = \Delta k = \Delta (v\gamma)$ then the Heisenberg Uncertainty Principle states:

$$\Delta x\, \Delta p \approx \hbar \quad \text{where} \quad \Delta p = \Delta (m\, v\gamma)$$

Therefore: $\Delta x\, \Delta (m\, v\gamma) \approx \boldsymbol{\hbar}$

Or, since the "ultimate" lower limit on Δx is of the order of the Compton wavelength, i.e. $\Delta x \approx \xi$ or $\hbar / mc$, we have:

$$m \approx \hbar / \underline{\Delta x}\, c$$

Where $\underline{\Delta x}$ denotes a lower limit to Δx. And it can be shown that the quantity $1/\xi$ possesses the additive and inertial properties of mass.

8.5. Bohm's Version of the Uncertainty Principle:

In Bohm's case, the Heisenberg relations are embodied in his theory as a limiting case *over a certain level of intervals of space and time*. However, the potential exists for the fields to be averaged over smaller intervals and hence, *subject to a greater degree of self-determination than is consistent with the Heisenberg principle*. As Bohm concludes[32]:

From this, it follows that our new theory is able to reproduce, in essence at least, one of the essential features of the quantum theory, i.e. Heisenberg's principle, and yet have a different content in new levels

Bohm is primarily concerned with the canonically conjugate field momentum, for which the associated coordinates, i.e. Δt, $\Delta\phi_k$ fluctuate at random. Thus, we have, according to Bohm:

$$\underline{\pi}_k = a\,(\Delta\phi_k\,/\Delta t)$$

Where a is a constant of proportionality, and $\Delta\phi_k$ is the fluctuation of the field coordinate. If then the field fluctuates in a random way the region over which it fluctuates is;

---

[32] Bohm, *op. cit.*,. 91.

$$(\delta \ \Delta\phi_k)^2 = b \ (\Delta t)$$

Taking the square root of both sides yields:

$$(\delta \ \Delta\phi_k) = b^{1/2} \ (\Delta t)^{1/2}$$

Bohm notes that π_k also fluctuates at random over the given range so:

$$\delta \pi_k = a \ b^{1/2} / \ (\Delta t)^{1/2}$$

Combining all the preceding results one finally gets a relation reflective of the Heisenberg principle, but time independent:

$$\delta \pi_k \ (\delta \ \Delta\phi_k) = ab$$

This is analogous to Heisenberg's principle, cf.

$$\delta p \ \delta q \leq \hbar$$

Where the product ab plays the same role as $\hbar$

End of Chapter Problems:

1) Consider the details of the Einstein 'box' thought experiment in which he attempted to out wit Bohr. Based on the hypothetic measured quantities given, how much uncertainty would be expected in the mass? In the weight? (Find the mass uncertainty in kg, and the weight in N.)

2)An Aspect-style experiment is carried out using the apparatus below:

```
P1 ↓| <------------|-------------> |↑ P2
    A1           D              A2
```

The result is found to be:

$$S = (A1,A2)I + (A1,A2)II + (A1,A2,)III + (A1,A2)IV =$$

$$2.65 \pm 0.10$$

Explain whether Bell's inequality was confirmed or not. If not, explain why not.

3) Explain what is meant by *causal covariant action at a distance* and specifically describe how it would appear in the experiment of (2). What would we expect to observe, say in the case of the polarizations of 2 gamma ray photons detected at A1 and A2?

4) According to the authors of *Gravitation and Spacetime* (p. 7) the upper limit on the mass of the photon is 10^{-59} g. Assume this to also be the rest mass in the photon's rest frame. Find its clock frequency according to the de Broglie- Bohm-Hiley pilot wave concept. Find its deBroglie wavelength if it's traveling at $v = c$. Compare this to the Compton wavelength.

5) Consider a theoretical wavelength, $\lambda = 2\pi / \Delta k$, where Δk is the expected width of the wave packet. If $\Delta k = 0.5$ nm, and $x = 1$ nm with $x_0 = 0.5$ nm, then compute the E-field amplitude: E_z Check the limits of the value you obtain using the prescription in the text.

6) In a particular experiment to test Bohmian quantum mechanics on a computer, the uncertainty in

one input turns out to be: $\Delta t = 10^{-39}$ s and in the other, $\Delta\phi_k = 10^{-51}$ m. From this data, find the quantity b. Then compose a form of the Uncertainty principle and obtain the product ab. (Where ab plays the same role as $\hbar$ in the conventional form of the Uncertainty principle.)

Comment on how your product ab compares to $\hbar$.

Then show from this that the deBroglie wavelength can be written: $\lambda_D = (h / m_0 c)$

Selected Solutions To End Chapter Problems :

Chapter I Solutions:

1)Given $x' = 1/a \, (x - vt)$ and $t' = 1/a \, (t - vx/c^2)$,

Then: $x' = x/a - vt/a$ and $t' = t/a - vx/ac^2$

and: $x' + vt/a = x/a$ and $t' + vx/ac^2 = t/a$

so:

$x = a(x' + vt/a)$ and $t' = a(t' + vx/ac^2)$

finally: $x = a(x' + vt)$ and $t = a(t' + vx/c^2)$

2)We have: $x' = 60m$, $t' = 8 \times 10^{-8}$ s and $y' = y$, $z' = z$

$v = 0.6c = 1.8 \times 10^8$ m/s

Then:

$x = [60m + (1.8 \times 10^8 \text{ m/s})(8 \times 10^{-8} \text{ s})]/ (0.64)^{\frac{1}{2}}$

$x = [60m + 14.4m]/ 0.8 = 74.4m/0.8 = 93m$

and $t =$

$[(8 \times 10^{-8} \text{ s})+ (1.8 \times 10^8 \text{ ms}^{-1})(60m)/(3 \times 10^8 \text{ ms}^{-1})^2/0.8$

$t = 2.5 \times 10^{-7}$ s$/ 0.8 = 2.33 \times 10^{-7}$ s

The space time coordinates are: (93 m, 2.33×10^{-7} s)

3)The problem requires no relative motion defined

specifically in the x-direction so the equations:

$$t = t' + x'v/c^2/(1 - v^2/c^2)^{1/2}$$

and:

$$t' = t - xv/c^2/(1 - v^2/c^2)^{1/2}$$

are immediately simplified by the terms in x becoming zero, so:

$$t = t'/(1 - v^2/c^2)^{1/2}$$

and $t' = t /(1 - v^2/c^2)^{1/2}$

Here: t = time passage on Earth clock

and t' = time passage on astronaut's clock

For t = 1 Earth year = 365 ¼ days:

$$t' = (365 \tfrac{1}{4} \text{ days})/ [1 - (0.9c)^2/c^2]^{1/2}$$

$$= (365 \tfrac{1}{4} \text{ days})/(1 - 0.81)^{1/2}$$

$$t' = (365 \tfrac{1}{4} \text{ days})/0.436 = 837.7 \text{ days}$$

This is the time elapsed on the astronaut's clock when the Earth has made one revolution equal to 365 ¼ days. In other words, each of his days is roughly equal to 2.29 Earth days. Hence, his clock is obviously running *slower* than the Earth clock.

A more intuitive way to look at the result would be in terms of the time transformation:

$t = t'/ (1 - v^2/c^2)^{1/2}$

- or asking how much time elapses on an Earth clock for each year elapsed on the astronaut's? The result will be found to be 2.29 years, or in other words his Earth counterparts are aging 2.29x faster.

4) We need to obtain t, for which:

$t = t'/ [1 - v^2/c^2]^{1/2}$

where v = 0.95c, so:

$t = 3.0 \text{ s}/ [1 - (0.95c)^2/c^2]^{1/2} = (3.0s) 1/ [0.0975]^{1/2}$

and t = (3.0 s)(3.2) = 9.6s

5) Given L = 11m and λ = 5.9 x 10^{-7} m

limit for lowest resolution: 2 Δ d = 0.005 fringe

Then:

d = (0.005 fringe)/ 2 = 0.0025 fringe

and since: Δ d = $Lv^2/ (c^2)$

The upper limit on the Earth's velocity is:

$v = \{(\Delta d)(\lambda) c^2)/ L\}^{1/2}$

$v = \{(0.0025)(5.9 \times 10^{-7} \text{ m})(3 \times 10^{8} \text{ ms}^{-1})/11 \text{ m}\}^{1/2}$

$= 3.5 \text{ km/s}$

6)We are going along (parallel) or opposed (anti-parallel) to, the "ether wind" direction from the diagram and this is *horizontal* so designate it by direction x, e.g.

t(total x)= L/ (c + v) + L/(c - v)

The common denominator is:

$(c - v)(c + v) = c^2 - cv + cv - v^2 = c^2 - v^2$

Then t(total x) = 2Lc/ $(c^2 - v^2)$

t(total x) = 2L/c $(1 - v^2/c^2)$-1

b) The time consumed for "a half-trip".

half trip time = (t(total x))/2 = ½ [2Lc/ $(c^2 - v^2)$]

(t(total x))/2 = Lc/ $(c^2 - v^2)$

But recall that the interferometer has parallel and perpendicular components so that total time registering both is:

T(total x+y) = t1 + t2 where:

t1 = 2Lc/ $(c^2 - v^2)$

$t2 = 2L/ (c^2 - v^2)^{1/2}$

So:

$t1 + t2 = 2Lc/ (c^2 - v^2) + 2L/ (c^2 - v^2)^{1/2}$

$= 2L \{c + (c^2 - v^2)^{1/2})/ (c^2 - v^2)\}$

so $T/2 = L\{c + (c^2 - v^2)^{1/2})/ (c^2 - v^2)\}$

c) Find the time consumed for a round trip.

Round trip is $2(t1 + t2)$

$= 2(2L\{c + (c^2 - v^2)^{1/2})/ (c^2 - v^2)\})$

$= 4L\{c + (c^2 - v^2)^{1/2})/ (c^2 - v^2)\}$

d) Add the two "*half trips*" and what do you obtain?

We get:

$2 (T/2) = 2L\{c + (c^2 - v^2)^{1/2})/ (c^2 - v^2)\}$

e) Why does this not agree with the value obtained for (a)?

It doesn't agree because each summing (or halving) as computed above takes into account *differing directional components* of velocity with respect to the interferometer. These need not be equal!
7) We have: $t = t'/ [1 - v^2/c^2]^{1/2}$

But the proper time is defined such that:

$t' = t/2$ or $t'/t = \frac{1}{2}$

Then:

$[1 - v^2/c^2] = (t'/t)^2$

and:

$v^2/c^2 = 1 - (t'/t)^2$

$v^2 = c^2[1 - (t'/t)^2]$

so:

$v = c[1 - (t'/t)^2]^{\frac{1}{2}} = c[1 - 0.5^2]^{\frac{1}{2}} = c[0.75]^{\frac{1}{2}} = 0.866c$

8) The proper time $t' = 3600$ s for the Jumbo jet

Since $v = 300$ m/s $= (10^{-6})$ c and hence $v/c \ll 1$ we need the form: $t = t'/[1 + v^2/2c^2]$ $t = 3600s/[1 + (10^{-12})c^2/2c^2]$ Since the numerator is only slightly larger than 1, the time t will be:

3600 s/(1.000000000001)= 3600.0000000018 = $3600 + 1.8 \times 10^{-9}$ s or slightly longer than one hour.

9)The proper time t' applies to the muon's reference frame.

So: $t = t'/[1 - v^2/c^2]^{\frac{1}{2}}$ and $t' = t[1 - v^2/c^2]^{\frac{1}{2}}$

where $v = 0.99$ c and $v^2 = (0.99c)^2 = 0.98c^2$

Then: $t' = t [1 - 0.98c^2/c^2]^{1/2} = t [0.02]^{1/2} = t(0.14)$

recall distance travelled = 4.6 km = 4600 m

To get t' we need to find t first, e.g.

$t = 4600m/ (2.97 \times 10^8 m/s) = 1.55 \times 10^{-5} s$

Then: $t' = (1.55 \times 10^{-5} s) (0.14) = 2.1 \times 10^{-6} s$

b) The distance traveled in its frame is just the proper length, L' so:

$L' = 4600 m [1 - v^2/c^2]^{1/2} = 4600m (0.02)^{1/2}$

$L' = 4600 m (0.14) = 644 m$

11) Let t_A be the time on the astronauts' clock and t_E be the time recorded on an Earth-based clock.

Then, we have $t_E = 4.4$ yrs.

And:

$t_A = t_E [1 - v^2/c^2]^{1/2}$

$t_A = (4.4 \text{ yrs.}) [1 - (0.95c)^2/c^2]^{1/2} = 1.37 \text{ yrs.}$

(b) Since we know: $t_A = 1.37$ yrs.

then the distance $D_A = (0.95c) (1.37 \text{ yrs}) = 1.31$ Ly

12) In this case, $v = c = 3 \times 10^8$ m/s

$d = v/H = (3 \times 10^8 \text{ m/s})/ (2.26 \times 10^{-18} \text{ s}^{-1})$

$d = 1.32 \times 10^{26} \text{ m}$

Converting to light years:

$d = (1.32 \times 10^{26} \text{ m})/ (9.5 \times 10^{15} \text{ m /Ly}) = 1.4 \times 10^{10} \text{ Ly}$

b) Would it be observable from Earth?

Given that modern telescopes can penetrate to about 1.8×10^{10} Ly, the galaxy should easily be observable to the Hubble but might be more problematical for land-based scopes.

14) We know the recessional velocity $v = 6 \times 10^4 \text{ km/s}$

By Hubble's law: $v = Hd$ so the distance $d = v/H$

Then, attending to the proper units for v, H:

$d = (6 \times 10^7 \text{ m/s})/(2.26 \times 10^{-18} \text{ s}^{-1})= 2.6 \times 10^{25} \text{ m}$

and $d = (2.6 \times 10^{25} \text{ m})/(9.5 \times 10^{15} \text{ m /Ly}) = 2.8 \times 10^9 \text{ Ly}$

(b) $z = v/c = (6 \times 10^7 \text{ m/s})/(3 \times 10^8 \text{ m/s}) = 0.2$

(c) $T = d/v = (2.6 \times 10^{25} \text{ m})/(6 \times 10^7 \text{ m/s})$
$= 4.3 \times 10^{17} \text{ s}$

But 1 yr. $= 3.15 \times 10^7 \text{ s}$

so $T = (4.3 \times 10^{17} \text{ s})/(3.15 \times 10^7 \text{ s/ yr})$

$T = 1.36 \times 10^{10}$ years, or 13.6 billion years

15) Let λ_o be the normal wavelength = 1200 Å and λ be *the red-shifted* value.

We know v = 0.8c so we must use the modified Doppler version, viz.

$\lambda/\lambda_o = (1 - v/c)^{1/2} /(1 + v/c)^{1/2}$

$\lambda/\lambda_o = (1 + 0.8)^{1/2}/ (1 - 0.8)^{1/2} = (1.8/0.2)^{1/2}$

$\lambda/\lambda_o = \sqrt{9} = 3$

then:

$\lambda = 3 \lambda_o = 3 (1200 \text{ Å}) = 3600 \text{ Å}$

(b) The red shift of the quasar is found from:

$1 + z = (1 + v/c)^{1/2} /(1 - v/c)^{1/2}$

$1 + z = (1.8/0.2)^{1/2} = \sqrt{9} = 3$

Then: z = 3 - 1 = 2

c) The *corrected velocity,*

$v = c [(z^2 + 2z) / (z^2 + 2z + 2)]$

Then: v = c[4 + 4]/ [4 + 4 + 2] = 8c/10 = 4c/5 = 0.8c

16) (a) by *the work -energy theorem:*

224

W = K(f) - K(i)

or:

$W = m_o \, c^2 / \, [(1 - u^2/c^2)^{1/2}] - m_o \, c^2$

where $m_o = 10^8$ kg and $u = 0.6c$

Then:

$W = (10^8 \text{ kg})c^2 / \, [(1 - (0.6c)^2/c^2)^{1/2}]$

$W = (10^8 \text{ kg})c^2 / \, ([1 - 0.36]^{1/2}) = (10^8 \text{ kg})c^2 / \, ([0.64]^{1/2})$

$W = (10^8 \text{ kg})c^2 / \, (0.8) = 1.25 \times 10^8 \text{ kg}(c^2) = 1.12 \times 10^{25}$ J

(b) According to theory, equal amounts (masses) of matter and antimatter are required for complete annihilation and total conversion of the initial masses into energy. The energy needed (as shown in (a)) is 1.12×10^{25} J. Thus: $m(a) = m(m) = 6.25 \times 10^7$ kg.

(17) The Cr 55 nucleus has a mass of 54.9279 u and the Mn 55 nucleus has a mass of 54.9244u, hence, the mass difference is:

$\Delta m = 54.9279 \text{ u} - 54.9244u = 0.0035$ u

(b) The maximum kinetic energy of the emitted electrons can be found using the mass defect.

The mass defect (from (a)) is 0.0035u so:

$E = (\Delta m) c^2 = (0.0035u)(931 \text{ MeV}/u) = 3.25 \text{ MeV}$

By the work-energy theorem:

$W = K(f) - K(i)$

$K(i) = 0 \ (u1 = 0)$

$K(f) = m_0 c^2 / [(1 - u^2/c^2)^{1/2}] = 3.25 \text{ MeV}$

where we need to find u.

Using a table one finds the rest energy of the electron = 0.511 MeV

and $K(f) = (3.25/ 0.511) = 6.36x$ the rest mass

So:

$E = 6.36mc^2 = mc^2 / [(1 - u^2/c^2)^{1/2}]$

$6.36 = 1/ [(1 - u^2/c^2)^{1/2}]$ or:

$(1 - u^2/c^2) = 1/(6.36)^2 = 1/40.44$

or:

$u^2/c^2 = 1 - 0.0024 = 0.9976$

Therefore: $u = [0.9976c^2]^{1/2} = 0.9987c$

(20) By the work-energy theorem:

$W = K(f) - K(i)$

where K(i) is just the initial rest energy, or $m_0 c^2$

Then $W = mc^2 / [(1 - u^2/c^2)^{1/2}] - m_0 c^2 =$

(total energy - rest energy)

Then, the total energy E:

$= mc^2 / [(1 - u^2/c^2)^{1/2})^{1/2}] = mc^2 / [(1 - 0.25)^{1/2}] = 1.16mc^2$

Then *the excess* $(E - E' = E_k)$ is that required from the fuel, or:

$E_k = (1.16mc^2 - mc^2 = 0.16mc^2)$

Mass of nuclear fuel - call it m'- is then related to E_k by:

$E_k = m'c^2 = 0.16m'c^2$ or

$(m'/m) = 0.16$ or $m' = 0.16m$

In other words, the nuclear fuel mass(m') needs to be at least 16% of the total initial mass of the rocket. So, if the rocket's mass is 100,000 kg then the nuclear fuel must be at least (0.16)(100,000 kg) = 16, 000 kg.

For time dilation:

$t' = t [1 - v^2/c^2]^{1/2}$

again, $v = 0.5$ c so:

$t' = t [1 - 0.25]^{1/2} = t [0.75]^{1/2} = 0.866t$

Now, there are 9.5×10^{15} meters per light year

Therefore the distance to Proxima Centauri is:

$D = (4.2 \text{ Ly}) \times (9.5 \times 10^{15} \text{ m/Ly}) = 4 \times 10^{16} \text{ m}$

Or: $D = 4 \times 10^{13} \text{ km}$

The time required is:

$(0.866/0.500) \times (4 \times 10^{13} \text{ km}) / (300,000 \text{ km/s}]$

$= 1.73 \times (4.22 \text{ yrs}) = 7.3 \text{ years}$

Since a generation is generally figured as 40 years, the time factor will not be an issue

(21) The normal position of the line is that in the solar spectrum since there is no appreciable radial velocity of the Sun with respect to Earth. Thus: $\Delta\lambda = (\lambda - \lambda_o)$

Where: $\lambda_o = 3968.49$ Å and $\lambda = 3968.20$ Å

Therefore:

$(\lambda - \lambda_o) = (3968.20\text{Å} - 3968.49 \text{ Å}) = -0.29\text{Å}$

The negative sign denotes motion toward the observer. Hence, approaching.

Since: $(\Delta\lambda / \lambda) = v/c$

The velocity $v = c (\Delta\lambda / \lambda) = c (-0.29\text{Å} / 3968.49 \text{ Å})$

$c = 3 \times 10^8 \text{ ms}^{-1}$

so: $v = 3 \times 10^8 \text{ ms}^{-1} (7.3 \times 10^{-5}) = 21\,900 \text{ ms}^{-1}$

Chapter II Selected Solutions:

1) a) The decay constant $\lambda = 0.693/ T_{1/2} =$

$0.693/ 883612800s$

Or: $\lambda = 7.84 \times 10^{-10}$

b) The activity A after 1 hour (3600 s) is given by:

$A = A_0 \exp(-\lambda t) = 1.1 \times 10^{10}$ decays/sec $(\exp(-\lambda t))$

Where $(\lambda t) = (7.84 \times 10^{-10})(3600 \text{ s}) =$

2.82×10^{-6}

Then:

$A_0 \exp(-\lambda t) = 1.1 \times 10^{10}$ /s $[\exp(\mathbf{-2.82 \times 10^{-6}})] =$

1.1×10^{10} /s $[0.99999] = 1.1 \times 10^{10}$ decays/sec

After two hours:

$(\lambda t) = (\mathbf{7.84 \times 10^{-10}})(7200 \text{ s}) = \mathbf{5.64 \times 10^{-6}}$

Then: $A_0 \exp(-\lambda t) =$

1.1×10^{10} /s $[\exp(\mathbf{-5.64 \times 10^{-6}})] = 1.1 \times 10^{10}$ /s

c) After 49 years, or t = 1546322400 s

$(\lambda t) = (7.84 \times 10^{-10})(1546322400 \text{ s}) = 1.21$

Then: $A_o \exp(-\lambda t) = 1.1 \times 10^{10}$ /s [exp (-1.21)]

So; $A = (1.1 \times 10^{10} \text{ /s})(0.2975) = 3.2 \times 10^9$ decays/sec

d)The number of nuclei is just: $N = N_o \ exp(-\lambda t)$,

So: $N = (2.5 \times 10^{17})(0.2975) = 7.4 \times 10^{16}$

e) We use: $N / N_o = exp(-\lambda t)$ where $N = 1.5 \times 10^{10}$

Then: $\log_e N - \log e \ N_o = -\lambda t$ or

$t = (\log_e N_o - 23.43) - \log_e \lambda$

(4) Given: $dN/dt = -lN$

So that: $\lambda = 1/N \ | \ dN/dt \ | = 10^{-15}(6.00 \times 10^{11}\text{/s})$

$\lambda = 6.00 \times 10^{-4} \text{s}^{-1}$

$A = -\lambda N = -(6.00 \times 10^{-4}\text{s}^{-1})(10^{15}) = -6.00 \times 10^{11}\text{/s}$ i.e.

6.00×10^{11} *decays per second*

c) Half life: $T_{1/2} = \ln 2/\lambda = 0.693/\lambda =$

$0.693/ (6.00 \times 10^{-4} s^{-1})$

$= 1160$ s (or 19.3 minutes)

5) The energies for the 1st, 2nd and 3rd Balmer transitions will be, respectively:

1st) $E_{(n=5 \text{ to } n = 2)} = -13.6 (1/ 5^2 - 1/ 2^2)$

2nd) $E_{(n=4 \text{ to } n = 2)} = -13.6 (1/ 4^2 - 1/ 2^2)$

3rd) $E_{(n=3 \text{ to } n = 2)} = -13.6 (1/ 3^2 - 1/ 2^2)$

Take differences between energy levels for Balmer lines:

Balmer α line (called H- alpha):

$E3 - E2 = -13.6$ eV $(1/ 3^2 - 1/ 2^2) = -$

13.6 eV$(1/9 - \frac{1}{4}) = -13.6$ eV $(-5/ 36) = 1.88$ eV

Now, 1 eV $= 1.6 \times 10^{-19}$ J so:

So: $E3 - E2 = 1.88$ eV $(1.6 \times 10^{-19}$ J $/$eV$) =$

3.02×10^{-19} J

From this, the wavelength of the photon emitted can be found. Since $E = hf = h (c/ \lambda)$:

$\lambda = hc/ (E3 - E2)$

$\lambda =$ (6.626069 x 10^{-34} J-s)(3 x 10^8 m/s)/ (3.02 x 10^{-19} J)

$\lambda =$ 6.56 x 10^{-7} m

Balmer β line (called H β):

$E4 - E2 = -13.6$ eV $(1/ 4^2 - 1/ 2^2)$

$= -13.6$ eV$(1/16 - ¼) = -13.6$ eV $(-3/16) =$ 2.55 eV

1 eV = 1.6 x 10^{-19} J so:

So: $E4 - E2 =$ 2.55 eV (1.6 x 10^{-19} J /eV) = 4.08 x 10^{-19} J

As before, the wavelength of the photon emitted is:

$\lambda =$ hc/ $(E4 - E2) =$

(6.626069 x 10^{-34} J-s)(3 x 10^8 m/s)/ (4.08 x 10^{-19} J)

$\lambda =$ 4.47 x 10^{-7} m

Balmer γ line (called Hγ):

$E5 - E2 = -13.6$ eV $(1/ 5^2 - 1/ 2^2)$

$= -13.6$ eV$(1/25 - ¼) = -13.6$ eV $(-21/100) =$ 3.4 eV

1 eV = 1.6 x 10^{-19} J so:

So: $E5 - E2 = 3.4$ eV $(1.6 \times 10^{-19}$ J /eV$) = 5.44 \times 10^{-19}$ J

As before, the wavelength of the photon emitted is inversely proportional to the difference between energy levels:

$\lambda = hc/ (E5 - E2) =$

$(6.626069 \times 10^{-34}$ J-s$)(3 \times 10^8$ m/s$)/ (5.44 \times 10^{-19}$ J $)$

$\lambda = 3.63 \times 10^{-7}$ m

(7) $K_{max} = 0.57$ eV and photo-electrons dislodged from a metal surface by incident radiation with $\lambda = 500$ nm.

The incident energy $E = hf = h c/\lambda = (6.62 \times 10^{-34}$ Js$) (3 \times 10^8$ ms$^{-1})/ (500 \times 10^{-9}$ m$)$

$h c/ \lambda = 3.97 \times 10^{-19}$ J

$\mathbf{K_{max}} = 0.57$ eV $(1.6 \times 10^{-19}$ J/ eV$) = 0.91 \times 10^{-19}$ J

Therefore:

$\mathbf{K_{max}} = 0.91 \times 10^{-19}$ J

$= [3.97 \times 10^{-19}$ J $- \phi]$

So, the work function is: $\phi = [3.97 \times 10^{-19}$ J $- 0.91 \times 10^{-19}$ J$] = 3.06 \times 10^{-19}$ J

To get in eV: $(3.06 \times 10^{-19} \text{ J})/ (1.6 \times 10^{-19} \text{ J}/ \text{eV}) = 1.91 \text{ eV}$

The stopping potential in volts:

$V_s =$ hf/ e - ϕ/e

where the slope h/e = 4.13×10^{-15} Js/C

The frequency $f = c/\lambda$ =

$(3 \times 10^8 \text{ ms}^{-1})/ (500 \times 10^{-9} \text{ m}) = 6 \times 10^{14} \text{ c/s}$

Then:

$V_s = 4.13 \times 10^{-15}$ J s/C $(6 \times 10^{14} \text{ s}^{-1})$ - $(3.06 \times 10^{-19}$ J)/ $(1.6 \times 10^{-19}$ C)

$V_s = 2.47 \text{ V} - 1.91 \text{ V} = 0.56 \text{ V}$

(4) (a): we note that for a multi-electron atom we have for the Bohr –atom energy:

$E_n = -13.6 (Z_{eff})^2/ n^2$ where: $Z_{eff} = (Z - 10)$

Note that Z_{eff} is the atomic number (Z =11) minus the number of electrons between the nucleus and the electron being considered. Since 10 electrons separate via ionization the last remaining sodium electron (e.g. the electron that occupies the 3s sub-level of the n= 3 principal level (electron arrangement for sodium is: ($1s^2$, $2s^2$, $2p^6$, $3s^1$)).

Then (given the 1st line corresponds to n = 3):

E_n (eV) $= -13.6 (Z-10)^2 / \mathbf{(3)}^2$

$E_n = -13.6$ eV $(1)/9 = 13.6$ eV$/9 = 1.51$ eV

Now, the energy is related to the wavelength (λ) in nm by:

$E\lambda = 1.99 \times 10^{-16}$ J nm

To get the energy in Joules, we use the fact that:

1 eV $= 1.6 \times 10^{-19}$J so 1.51 eV $= 1.51$ eV$(1.6 \times 10^{-19}$J /eV)

$E = 2.41 \times 10^{-19}$ J Then:

$\lambda = 1.99 \times 10^{-16}$ J nm $/ 2.41 \times 10^{-19}$ J $= 825$ nm

   (b) The voltage of 10^4 V leads to an energy:

$E = qV = (1.6 \times 10^{-19}C)(10^4$ V$) = 1.6 \times 10^{-15}$ J

(9) We need the possible values of s', ℓ' and j' for an atomic configuration with 2 optically active electrons with quantum numbers: $\ell 1 = 2$, s1 = ½, $\ell 2 = 3$, s2 = ½.

So $\ell 1 = 2$ and $\ell 2 = 3$ therefore: $\ell 1 + \ell 2 = 2 + 3 = 5$.

Meanwhile, S can be defined by only one value of s, or s = ½

The possible j-values are:

$j = \ell + s = \ell 1 + s = 2 + ½ = 5/2$

$j = \ell_2 + s = 3 + \frac{1}{2} = 7/2$

$j = \ell - s = \ell_1 - s = 2 - \frac{1}{2} = 3/2$

$j = \ell_2 - s = 3 - \frac{1}{2} = 5/2$

So: $j' = 7/2,\ 5/2,\ 3/2$

$s' = \frac{1}{2}$ and $-\frac{1}{2}$

10) Draw a schematic energy diagram for the 2p 3s configuration for the carbon atom (^{12} C $_6$) and label each level with spectroscopic notation.

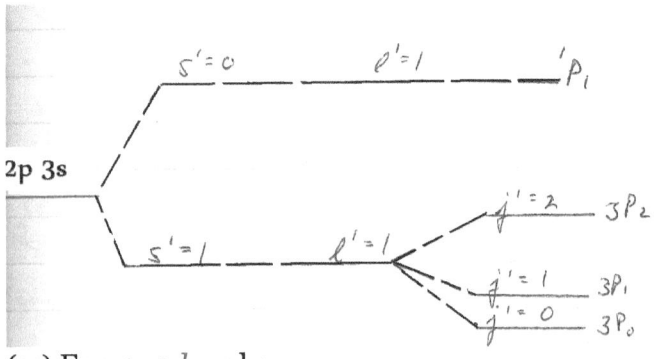

(11) For *4s 3d* we have:

$\ell_1 = 2$, $s_1 = 1/2$, $\ell_1 = 0$ and $s_1 = 1/2$. Then for the maximum value:

$\ell_1 + \ell_2 = 2 + 0 = 2$

and: $s = s_1 + s_2 = 1/2 + 1/2 = 1$

The *lowest* energy level is then:

236

4s 3d ($3D1$)

Since 2s' + 1 = 3, leading to minimum:

j' = [s' - ℓ '] = 1.

Using the assorted combinations, for ℓ '= 0 and ℓ ' = 2, to get the respective j' values (in combination with s' = 0) and then further for s' = 1, we arrive at the energy configuration diagram below:

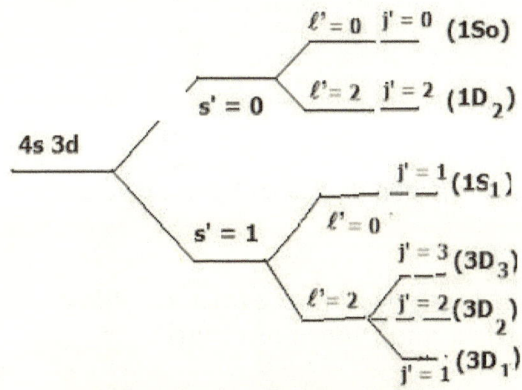

Schematic Energy Level diagram for 4s 3d: Note the spectroscopic notation is given in bracket at the far right.

(12)[Kr] $4d^9 5s^1$ state has a higher energy than the [Kr] $4d^9$ state, since in the former we have the principal quantum number n = 5 which implies greater quantized energy (e.g. E_n = -13.6/ n^2). Therefore, the higher 'n' value means the energy state E_n = E_5 is less negative and hence, greater.

(13)For a two-electron atom for which the orbital angular momentum quantum numbers are $\ell_1 = 3$ and $\ell_2 = 2$:

$\ell_1 = 3$ and $\ell_2 = 2$ so $\ell_1 + \ell_2 = 2 + 3 = 5$

Therefore, the possible values of L will be found from letting $\ell_1 = 3$ and adding each next descending value of m_ℓ from 2, to 1, to 0, to -1, to -2:

$(3) + 1 = 4$

$(3) + 0 = 3$

$(3) + (-1) = 2$

$(3) + (-2) = 1$

So the total angular momentum L can have the values: 5, 4, 3, 2 and 1.

The *f electron* has $\ell = 3$ so that the *total angular momentum quantum number* possibilities are:

$j = \ell + \tfrac{1}{2}, \quad \ell - \tfrac{1}{2}$

Then: $j = 7/2, \ 5/2$

Chapter III Selected Solutions:

(1) We use $P_{ab} = \int {}^b_a \ |\psi|^2 \ dx$

Then: $P_{ab} = \int {}^{L/4}_0 \ (\sqrt{2}/ \sqrt{L})^2 \ \sin^2 (2\pi x/L) \ dx$

238

Let $\theta = 2\pi x/L$ and use: $\sin^2 \theta = \frac{1}{2}(1 - \cos 2\theta)$

$P_{ab} = 2/L \int_0^{L/4} \frac{1}{2}[1 - \cos(4\pi x/L)]\, dx$

$P_{ab} = 2/L\,[\frac{1}{2}\int_0^{L/4} dx - \int_0^{L/4} \cos(4\pi x/L)]\, dx$

And: $P_{ab} = x/L - 1/2\pi \sin(4\pi x/L)]_0^{L/4}$

$P_{ab} = \frac{1}{4} - 0 = 0.25$

Or a 1 in 4 probability of being found in that region.

(3) $\psi = A \sin(n\pi x/L)$

And: $\psi^2 = A^2 \sin^2(n\pi x/L)$

For normalization we require:

$1 = = \int_0^L P(x)\, dx = A^2 \int_0^L \sin^2(n\pi x/L)\, dx$

Whence:

$A^2 (L/n\pi)\,[u/2 - \sin 2u/4]_0^{n\pi} = 1$

$A^2 (L/n\pi)(n\pi/2) = 1$

Or: $A^2 = 2/L$

Therefore: $A = \sqrt{2}/\sqrt{L}$

(4) From the information and the Heisenberg Uncertainty Principle:

$\Delta P x \geq h/ 2\pi = 1.05 \times 10^{-34}$ J-s

The uncertainty in the x-component is 0.5 angstrom where $1 A = 1.0 \times 10^{-10}$ m

Then: $\Delta x = 0.5A = 5.0 \times 10^{-9}$ cm

The uncertainty in the x-component of the momentum of the electron is:

$P \approx 1.05 \times 10^{-27}$ erg-s/ 5.0×10^{-9} cm $= 2.1 \times 10^{-24}$ kg m-s

Or, 2×10^{-19} g cm-s

(5) (a) In classical theory the probability that the particle would be in the region between x and x + L/3 is 0.33. Since no other information is given then it is equally probable that it's in any place in the region – so one third of the time will be spent over any given line segment, i.e. which is one-third the total length.

(b) For a quantum mechanical interpretation:

$\psi = A \sin 2kx$

$\sin 2kL = 0$ and k (wave number) $= n\pi/L$

$\psi = A \sin (n\pi x/L)$

$P(x) = \psi^2 = A^2 \sin^2 (n\pi x/L)$

For normalization:

$$1 = = \int_{L_0} P(x) \ dx = A^2 \int_{L_0} \sin^2 (n \pi x/L) \ dx$$

$$A^2 \ (L/ n \pi) \left[u/2 - \sin 2u/ 4 \right]_0^{n\pi} \quad = 1$$

$$A^2 \ (L/ n \pi) (n \pi/2) = 1$$

Or: $A^2 = 2/L$

Then the probability it will lie between 0 and $L/3$ is:

$$\int_{L/3_0} \sin^2 (n\pi x/L) \ dx$$

$$= 2/ n \pi \left[n \pi/ 6 - \tfrac{1}{4} (\sin 2 \pi/ 3)\right] =$$

$$2/ \pi \left[\pi/ 6 - \tfrac{1}{4} (\sin 2 \pi/ 3)\right]$$

$$= 0.19$$

(Since $n = 1$ is the lowest energy state)

(c) Since $n = 2$ is the second lowest state the probability the particle is between 0 and $L/3$ is:

$$(2/L) L/ 2 \pi \ \left[\pi/ 3 - \tfrac{1}{4} (\sin 4 \pi/ 3)\right] = 0.40$$

Chapter IV Selected Solutions:

2) From the information and the Heisenberg Uncertainty Principle:

$$\Delta P x \ \geq h/ 2\pi = 1.05 \times 10^{-34} \ \text{J-s}$$

The uncertainty in the x-component is 0.5 angstrom where $1 A = 1.0 \times 10^{-10}$ m

Then we have: $\Delta x = 0.5A = 5.0 \times 10^{-9}$ cm

The uncertainty in the x-component of the momentum of the electron is:

$P \approx 1.05 \times 10^{-27}$ erg-s$/ 5.0 \times 10^{-9}$ cm $= 2.1 \times 10^{-24}$ kg m-s

Or, 2×10^{-19} g cm-s

(3) (a)   In classical theory the probability that the particle would be in the region between x and $x + L/3$ is 0.33. Since no other information is given then it is equally probable that it's in any place in the region – so one third of the time will be spent over any given line segment, i.e. which is one-third the total length.

(b) For a quantum mechanical interpretation:

$\psi = A \sin 2kx$

$\sin 2kL = 0$ and k (wave number) $= n\pi/L$

$\psi = A \sin (n\pi x/L)$

$P(x) = \psi^2 = A^2 \sin^2 (n\pi x/L)$

For normalization:

$1 = = \int_0^L P(x)\ dx = A^2 \int_0^L \sin^2 (n\pi x/L)\ dx$

242

$A^2 (L/n\pi) [u/2 - \sin 2u/4]_0^{n\pi} = 1$

$A^2 (L/n\pi)(n\pi/2) = 1$

Or: $A^2 = 2/L$

Then the probability it will lie between 0 and L/3 is:

$\int_0^{L/3} \sin^2(n\pi x/L)\, dx$

$= 2/n\pi [n\pi/6 - \tfrac{1}{4}(\sin 2\pi/3)] =$

$2/\pi [\pi/6 - \tfrac{1}{4}(\sin 2\pi/3)]$

$= 0.19$

(Since n = 1 is the lowest energy state)

(c) Since n = 2 is the second lowest state the probability the particle is between 0 and L/3 is:

$(2/L) L/2\pi [\pi/3 - \tfrac{1}{4}(\sin 4\pi/3)] = 0.40$

---

(4) For the step potential, show that for the region x > 0   and E < V$_0$ With $K_1 = \sqrt{(2m(V_0 - E)}/\hbar$

The general solution of the appropriate Schrodinger equation is:

$\psi(x) = C \exp(K_2 x) + D \exp(-K_2 x)$

$d\psi(x) / dx = CK2 \exp(K2\ x) - D\ K2 \exp(-K2\ x)$

$d^2\psi(x) / dx^2 =$

$CK2^2 \exp(K2\ x) + D\ K2^2 \exp(-K2\ x)$

$= K2^2\ \psi(x) = \psi(x)\ [2m(V_0 - E) / \hbar^2]$

Which yields on substitution:

$-\hbar^2/ 2m\ \ [2m(V_0 - E) / \hbar^2]\ \psi(x)\ \ + V_0\ \psi(x) = E\psi(x)$

b) For the transmitted (T) and reflected(R) flux of the step potential,

$R = (K1 - K2)^2 / (K1 + K2)^2$

$T = 4K1K2 / (K1 + K2)^2$

$R + T =$

$(K1 - K2)^2 / (K1 + K2)^2\ +\ 4K1K2 / (K1 + K2)^2$

$= [K1^2 - 2K1K2 + K1^2\ +\ 4K1K2] / (K1 + K2)^2$

$=\ \ [K1^2\ +\ 2K1K2 + K1^2] / (K1 + K2)^2$

$=\ \ (K1 + K2)^2 / (K1 + K2)^2\ \ =\ \ 1$

6) Using the operators, $\mathbf{p}^\wedge r$ and $\ell^{\wedge 2}$

$= -\hbar^2/ \sin^2\theta\ [\ \sin\theta\ \partial / \partial\theta\ (\sin\theta\ \partial / \partial\theta) + \partial^2 / \partial\varphi^2\ ]$

We write out the full form of the Schrodinger equation for the hydrogen atom in spherical coordinates. Then obtain the final form such that $\mathbf{H}_{op} = -[E - V]\,\psi\,(r, \theta, \varphi)$

Replace m by the reduced mass,

$\mu = mM/\,m+M$

Write out the operator:

$\mathbf{H}_{op} = \mathbf{p}\char`\^\mathbf{r}^2/2m + \ell^{\wedge 2}/2mr^2 + V(r)$

And $\psi = \psi\,(\mathbf{r}, \boldsymbol{\theta}, \boldsymbol{\varphi})$

Then: $\mathbf{H}_{op}\ \psi = E\,\psi =$

$[\mathbf{p}\char`\^\mathbf{r}^2/2m + \ell^{\wedge 2}/2mr^2 + V(r)]\,\psi\,(\mathbf{r}, \boldsymbol{\theta}, \boldsymbol{\varphi})$

$= E\psi\,(\mathbf{r}, \boldsymbol{\theta}, \boldsymbol{\varphi})$

Where:

$\mathbf{p}\char`\^\mathbf{r}^2 = (-i\hbar\,1/r\,(\partial\,(r)/\,\partial r\,)^2 =$

$-\hbar^2\,(1/r^2\,\partial/\partial r\,(r^2\partial/\partial r\,)$

$\ell^{\wedge 2} =$

$-\hbar^2/\sin^2\theta\,[\sin\theta\,\partial/\partial\theta\,(\sin\theta\,\partial/\partial\theta) + \partial^2/\partial\varphi^2\,]$

Then we may write out the Schrodinger equation in spherical coordinates:

$- \hbar^2 / 2m \, (\partial / \partial r \ (r^2 \partial / \partial r\,) -$

$\hbar^2 / 2mr^2 \sin^2 \theta \, [\, \sin \theta \, \partial / \partial \theta \, (\sin \theta \, \partial / \partial \theta)$

$+ \, \partial^2 / \partial \varphi^2 \,] \, \psi \,(r, \theta, \varphi) + \, V(r) \, \psi \,(r, \theta, \varphi)$

Replace m by the reduced mass,

$\mu = mM / m+M$:

$- \hbar^2 / 2\mu \, [\, \partial / \partial r \ (r^2 \partial / \partial r\,) -$

$1 / \, r^2 \sin^2 \theta \quad \sin \theta \, \partial / \partial \theta \, (\sin \theta \, \partial / \partial \theta)$

$+ \, \partial^2 / \partial \varphi^2 \,] \, \psi \,(r, \theta, \varphi) + \, V(r) \, \psi \,(r, \theta, \varphi) = E\psi \,(r, \theta, \varphi)$

Which is the appropriate Schrodinger equation in spherical coordinates.

7)In the case of the first solution (i) we demand that the function be single-valued and continuous so that:

$\Phi(\varphi + 2\pi) \ = \Phi(\varphi)$

So that:

$\exp [\, i \, m_\ell \, (\varphi + 2\pi)] = \exp (i \, m_\ell \varphi)$

where m_ℓ is the magnetic quantum number. Then:

$\exp [\, i \, m_\ell \, (\varphi + 2\pi)] = \exp (i \, m_\ell \varphi) = \ 1$

By de Moivre's theorem:

$\cos 2\pi\, m_\ell + i \sin 2\pi\, m_\ell = 1$

But: $\sin 2\pi\, m_\ell = 0$ so: $\cos 2\pi\, m_\ell = 1$

Which requirement is satisfied provided the absolute value of m_ℓ has one of the following values:

$|\, m_\ell\,| = 0, 1, 2, 3, 4\ldots\ldots$

(11)We have for the harmonic oscillator Hamiltonian:

$\mathbf{H}^{\wedge} = \mathbf{p}^{\wedge 2}/\, 2m + m\omega^2\, x^2/\, 2$

And the kinetic energy operator: $\mathbf{p}^{\wedge} = i\,\hbar\; (\partial\,/\partial x)$.

It is useful to sketch the quantum potential at this point and also get the normalization condition. We have:

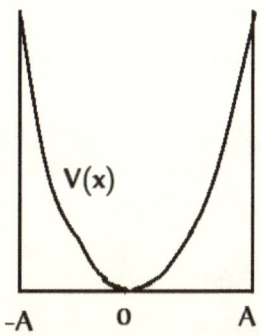

And since we know the oscillator (on account of tunneling can penetrate beyond the $-A \leq x \leq A$ limits of the potential then we need to use:

$\psi_0\,(y) = A \exp\,[-y^2/\,2]$ where $y = (m\omega/\,h)^{1/2}\,x$

To get the value of A the normalizing factor, we have:

$$\int_{-\infty}^{\infty} |\psi(y)|^2 \, dy = 1$$

Then:

$$|\psi(y)|^2 = [A \exp(-y^2/2)]^2 = A^2 \exp(-y^2)$$

So that: $A^2 \int_{-\infty}^{\infty} \exp(-y^2) \, dy = 1$

And we may use the well known integral:

$$\int_{-\infty}^{\infty} \exp(-y^2) \, dy = \sqrt{\pi}$$

Simplifying:

$A^2 \sqrt{\pi} = 1$ and $A = (1/\sqrt{\pi})^{1/2}$

We now modify the energy operator $\mathbf{p}^\wedge = i\hbar \, (\partial/\partial x)$.

Instead we use: $\mathbf{p}_y^\wedge = \mathbf{p}_y{}^{\wedge 2}/2m = (-i\hbar \sqrt{\alpha} \, \partial/\partial y)^2/2m$

$= \alpha \hbar^2/2m \, [\partial^2/\partial y^2]$

(Note: $\sqrt{\alpha} = m\omega/\hbar)^{1/2}$

The appropriate operator equation is then:

$\mathbf{H}_y{}^\wedge = (-i\hbar \sqrt{\alpha} \, \partial/\partial y)^2/2m + y \, (m\omega/h)^{-1/2}$

And Schrodinger's equation becomes – in terms of $\psi(y)$ and α:

$$\frac{d^2\psi}{dy^2} + (\alpha - y^2)\psi = 0$$

The acceptable solutions for the above are constrained by the condition:

$$\int_{-\infty}^{\infty} |\psi(y)|^2 \, dy = 1$$

Which shows that as $\psi \to 0$, $y \to \infty$

The properties of the given Schrodinger's equation are such that the normalization condition can only be fulfilled if:

$\alpha = 2n + 1$ and $n = 0, 1, 2, 3 \ldots$

Since $\sqrt{\alpha} = (m\omega/\hbar)^{1/2}$ and $\omega = 2\pi\nu$

But $E = \hbar\nu$ then: $\alpha = 2E/\hbar\nu$

The energy levels of the harmonic oscillator are:

$E_n = (n + \frac{1}{2})\hbar\nu$ $n = 0, 1, 2, 3$

The energy is thereby quantized in steps of $\hbar\nu$ with the zero point energy:

$E_0 = \hbar\nu/2$

Because now : $\alpha_n = 2E_n/\hbar\nu$

We see that each choice of an α_n leads to a different wave function. Each function then introduces a polynomial called a Hermite polynomial of the form:

H $_n$ (y) yielding either odd or even powers of y. Thus, the Hermite polynomial must be introduced as part of any solutions for the wave function.

The form for any nth wave function related to the harmonic oscillator is then:

$$\psi_n = \left(\frac{2m\nu}{\hbar}\right)^{1/4}(2^n n!)^{-1/2} H_n(y)e^{-y^2/2}$$

It is useful to sketch the quantum potential at this point and also get the normalization condition. We have for the potential:

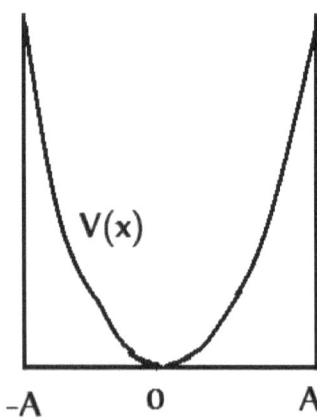

Chapter V Selected Solutions:

2) Appeal may be made to the "liquid drop" nuclear model. The diagram shown actually shows a constancy of E_b / A beyond the A »22 value and

beyond. I.e. if a correction is made at smaller A of this range for surface effects – analogous to the surface tension in a water droplet – and a correction is also made at larger A for Coulomb repulsion (of protons) then the main portion of the binding energy per nucleon curve (E_b / A vs. A) is found to be nearly constant.

Minus the liquid drop and Coulombic repulsion corrections noted above, the region of highest stability shown is simply the region for which E_b / A = max is effectively constant over:

$\approx 50 \leq A \leq 75$

 (Without applying either of the corrections.)

3) We have: $E_n = (h^2 \ p^{-2} / \ 8m \ L^2) \ (N/2)^2$

$= 4[(6.62 \times 10^{-34} \ \text{J-s})^2 \ (3.14)^2 \ / \ (8m \ L^2)$

Where L is obtained from: $R = r_o \ A^{1/3}$

$= (1.2 \times 10^{-15} \ \text{m}) \ (16)^{1/3} \ =$

$(1.2 \times 10^{-15} \ \text{m}) \ 2.5 = 3.0 \ \times 10^{-15} \ \text{m}$

Then: $E_n =$

$4[(6.62 \times 10^{-34} \ \text{J-s})^2 \ (3.14)^2 \ / \ [\ (8 \ (1.7 \times 10^{-27} \ \text{kg})(\ 3.0 \ \times 10^{-15} \ \text{m})^2]$

$E_n = \ 1.4 \times 10^{-10} \ \text{J}$

5)(a) The ratio of the radii is given by:

$R_1/R_2 = [r_o A_1^{1/3}] / [r_o A_2^{1/3}]$

Or: $R_1/R_2 = (A_1/A_2)^{1/3}$

Where: $A_1 = 4$ and $A_2 = 292$

Then:

$R_1/R_2 = (4/292)^{1/3}$

$= 0.239$

(b) The expression for the nuclear density, as a function of the Fermi energy is given by:

$r = [2M\, E_F / 3^{2/3}\, \pi^{4/3}\, \hbar^2]^{3/2}$

Then the ratio of densities, i.e. ρ_1/ρ_2 would be:

$\rho_1/\rho_2 =$

$[2M_1\, E_{1F}/ 3^{2/3}\, \pi^{4/3}\, \hbar^2]^{3/2} / [2M_2 E_{2F} / 3^{2/3}\, \pi^{4/3}\, \hbar^2]^{3/2}$

Simplifying:

$\rho_1/\rho_2 = [M_1\, E_{1F} / M_2\, E_{2F}]$

Thus, an estimate of the density ratio can be obtained by taking the ratio of the nucleon masses M_1 to M_2, and multiplying it by the ratio of the Fermi energies.

6) a) We need: $R = r_o A^{1/3}$ and: $A = 202$
Where: $r_o = 1.2 \times 10^{-15}$ m, so $D = 2R$

Then: $D_2 = 2 (1.2 \times 10^{-15} \text{ m}) (202)^{1/3} = 1.4 \times 10^{-14}$ m

And the diameter of hydrogen's nucleus is:

$D_1 = 2 \,(1.2 \times 10^{-15} \text{ m}) \,(1)^{1/3} = 1.2 \times 10^{-15} \text{ m}$

So the ratio (which yields how many times the element's nucleus is larger) is:

$D_2/ D_1 = 11.6$

b) The mass defect Δ M for this nucleus. We need M to get Δ M.

$80 \,(1.008142) + (202 - 80) \,[1.008982] = 203.747\text{u}$

$\Delta M = 201.\,970^* - 203.747 = -1.77 \text{ u}$

(* From table of atomic masses)

(c) *The binding energy* is then: $E_b = \Delta M \, c^2 = (931$ MeV/u)(1.77) = 1647.8 MeV

From this, the **binding energy per nucleon** is:

$E_b / A = 1647.8$ MeV/ 202 = 8.15 MeV / nucleon

7) From the information given, we have applicable to the gamma ray enegy:

$E \,(\gamma) = \text{hc}/\lambda = 6.7 \text{ keV}$

So: $6.7 \text{ keV} = (6.62 \times 10^{-34} \text{ J-s}) \,(3 \times 10^8 \text{ m/s}\,) / \lambda$

$\lambda = (6.62 \times 10^{-34} \text{ J-s}) \,(3 \times 10^8 \text{ m/s}\,) / 6.7 \text{ keV}$

Converting to consistent energy units, using 1 eV = 1.6 $\times 10^{-19}$ J:

$\lambda = (1.98 \times 10^{-26} \text{ J-m})/ (1.07 \times 10^{-16} \text{ J}) = 1.85 \times 10^{-10} \text{ m} = 185 \text{ nm}$

8) The 2nd part of the triple alpha fusion reaction is:

$^8\text{Be} + {}^4\text{He} \rightarrow {}^{12}\text{C} + \gamma + 7.4 \text{ MeV}$

$Q = [(8.00531 \text{ u} + 4.00260 \text{ u}) - 12.0000 \text{ u}] \text{ c}^2$

$Q = [12.00795 \text{ u} - 12.0000 \text{ u}] = 0.00795 \text{ u} (931 \text{ MeV/u}) = 7.4 \text{ MeV}$

The role (and value in energy) of the gamma ray photon can be obtained by using instead the value for carbon of 12.011 u and following the procedure presented in the chapter.

9) The luminosity is the same as the power or energy generated per unit time, thus:
$L = E/ t = 3.9 \times 10^{26} \text{ J/s}$

The ***energy*** delivered per second then is:

$3.9 \times 10^{26} \text{ J} = E = m \text{ c}^2$ so the mass converted to energy is:

$m = E/ c^2 = (3.9 \times 10^{26} \text{ J})/ (3 \times 10^8 \text{ m/s })^2$
$= 4.3 \times 10^9 \text{ kg}$

We have to assume the luminosity represents the

actual macroscopic mass converted into energy and is a faithful reflection of all the fusion reactions underlying the conversion.

13)(a) When ^{118} Sn $_{50}$ is bombarded with a proton the main fission fragments are: ^{24} Na $_{11}$ and ^{94} Zr $_{40}$ The excitation energy necessary for passage over the potential barrier is:

$\varepsilon > 3 k e^2 Z^2 / 5$

We have Z = 50 so that $\mathbf{Z^2}$ = 2500, then:

$\varepsilon = 3(1.44 \text{ MeV}/ \text{fm}) \; 2500/ 5 = 2160 \text{ MeV}$
(b) The fission reaction can be written:

^{118} Sn $_{50}$ $+ ^1 H_1 \rightarrow$ ^{24} Na $_{11}$ $+ ^{94}$ Zr $_{40}$
Therefore (Using atomic masses in physics formulary):

Q =
[(117.901606 u + 1.007825 u) - 24.99096u - 93. 907303 u] (931.5 MeV/u)

Q = [118.909431 - 118.898263] (931.5 MeV/u)

Q = [0.011168] (931.5 MeV/u) = 10. 4 MeV

15) u_I = A cos (ar) + B sin(ar)

Then: du_I/dr = - a A sin (ar) + a B cos(ar)

$d^2 u_I /dr^2$ = - a^2 A cos(ar) - a^2 B sin (ar) =

$- a^2 [A \cos(ar) + B \sin(ar)]$

Or: $d^2 u_I / dr^2 = - a^2 u_I$

Transposing:

$d^2 u_I / dr^2 + a^2 u_I = 0.$

Since $R = u/r$ the cosine solution must be discarded, lest we get an unwanted infinity. This leaves: $u_I = B \sin(ar)$

Chapter VI Selected Solutions

1) We first find the number of electrons per unit volume (N/V)

$N/V =$

6.020 x 10^{26} atoms/ kmol $(81$ kg/ m$^3)/$ 3 kg/ kmol

(Why 3 kg/ kmol in denominator? Because the atomic weight = 3)

Then: $N/V = 1.625$ x 10^{28} atoms

The Fermi Energy, $\varepsilon_F = \hbar^2 / 2m \ [3p^2 N/v]^{2/3}$

$\varepsilon_F = 6.81$ x 10^{-23} J

To get: v_F , note: $\varepsilon_F = \frac{1}{2} m (v_F)^2$

$v_F = [2 e \varepsilon_F / m]^{1/2} = 165 \text{ m/s}$

Finally, the Fermi temperature is found based on the fact it is tied to the energy (as the energy increases, the temperature increases. We have for the thermodynamic temperature:

Since $T_F = k_B t$

Then: $T_F = (\varepsilon_F)/ k_B =$

$(6.81 \quad \times 10^{-23} \text{ J}) / (1.38 \times 10^{-23} \text{ JK}^{-1}) = 4.93 \text{ K}$

2) We have:

$f(\varepsilon) = 1/ \{\exp (\mu - \varepsilon)/ \tau + 1\} =$

$[\exp (\mu - \varepsilon)/ \tau + 1]^{-1}$

$- \partial f / \partial \varepsilon =$

$- \{- [\exp (\mu - \varepsilon)/ \tau + 1]^{-2} \bullet 1/ \tau \; (\exp (\mu - \varepsilon)/ \tau)\}$

For $\varepsilon = \mu$:

$- \partial f / \partial \varepsilon = \exp (\varepsilon - \varepsilon)/ \tau + 1]^{-2} \bullet 1/ \tau \; (\exp (\varepsilon - \varepsilon)/ \tau)\}$

$= \exp 0/ \tau + 1]^{-2} \bullet 1/ \tau \; (\exp 0 / \tau)\} = 1/ \tau \; (2)^{-2}$

$- \partial f / \partial \varepsilon = 1/ 4 \tau = 1/ (4 k_B T) = (4 k_B T)^{-1}$

3)

If $\varepsilon = \mu + \delta$, then: $\delta = \varepsilon - \mu$

$f(\delta) = 1/[\exp(\varepsilon-\mu)/\tau + 1]$

But: $f(-\delta) = 1/[\exp - (\varepsilon-\mu)/\tau + 1]$

$= 1/[\exp(\mu-\varepsilon)/\tau + 1]$

And:

$1 - f(-\delta) = 1 - 1/[\exp(\mu-\varepsilon)/\tau + 1]$

$= [\exp(\mu-\varepsilon)/\tau]/[\exp(\mu-\varepsilon)/\tau + 1]$

$= \lambda \exp(-\varepsilon/\tau) / \lambda \exp(-\varepsilon/\tau) + 1$

Since: $\lambda = \exp(\mu/\tau)$

Multiply numerator and denominator by using: $[\exp(\varepsilon/\tau)]/$

$\lambda \exp(-\varepsilon/\tau)(\exp(\varepsilon/\tau)/\lambda) / [\lambda \exp(-\varepsilon/\tau) + 1](\exp(\varepsilon/\tau)/\lambda$

$= 1/[1 + 1/\lambda(\exp(\varepsilon/\tau)]$

Or: $1 - f(-\delta) = 1/[1 + \exp -\mu(\exp(\varepsilon/\tau)]$

$= 1/[\exp(\varepsilon-\mu)/\tau + 1]$

258

But: $f(-\delta) = 1/[\exp(\varepsilon-\mu)/\tau + 1]$

Therefore: $f(\delta) = 1 - f(-\delta)$

Chapter VII Selected Solutions

2) $\mathrm{Tr}(\mathbf{g}_{ik}) = 1 + \mathbf{r}^2 + \mathbf{r}^2 \sin\theta$

3) $\mathbf{x}_i = (2, 1, 4)$ and

$\mathbf{y}_j = (3, 7, -1)$

+

$$a_{ij} := \begin{pmatrix} 3 & 0 & 0 \\ 1 & 2 & 5 \\ -1 & 4 & 2 \end{pmatrix}$$

We obtain in turn:
(a)

$$a_{ij} \begin{pmatrix} 2 \\ 1 \\ 4 \end{pmatrix} = \begin{pmatrix} 6 \\ 24 \\ 10 \end{pmatrix}$$

And:

(b)

$$a_{ij} - \frac{2}{3} \cdot \delta_{ij} = \begin{pmatrix} 2.333 & 0 & 0 \\ 1 & 1.333 & 5 \\ -1 & 4 & 1.333 \end{pmatrix}$$

4) We use the same a_{ij} as for (3)

And:

$$\delta_{ij} := \begin{pmatrix} 1 & 0 & 0 \\ 0 & 1 & 0 \\ 0 & 0 & 1 \end{pmatrix}$$

Then:

$$a_{ij} \cdot \delta_{ij} = \begin{pmatrix} 2 & 0 & 3 \\ 5 & 1 & 2 \\ 4 & 5 & 7 \end{pmatrix} \begin{pmatrix} 1 & 0 & 0 \\ 0 & 1 & 0 \\ 0 & 0 & 1 \end{pmatrix}$$

Therefore:

$$a_{ij} \cdot \delta_{ij} = \begin{pmatrix} 2 & 0 & 3 \\ 5 & 1 & 2 \\ 4 & 5 & 7 \end{pmatrix}$$

$$\text{Tr}(a_{ij} \cdot \delta_{ij}) = 2 + 1 + 7 = 10$$

5) The *anti-symmetric part* follows from the basic definition:

$$a_{ji} := \begin{pmatrix} 0 & 3 & -4 \\ -3 & 0 & -2 \\ 4 & 2 & 0 \end{pmatrix}$$

The symmetric part is the difference: $a_{ij} - a_{ji} =$

$$\begin{pmatrix} 2 & 3 & 2 \\ 5 & 7 & -2 \\ 4 & -4 & 0 \end{pmatrix} - \begin{pmatrix} 0 & 3 & -4 \\ -3 & 0 & -2 \\ 4 & 2 & 0 \end{pmatrix} = \begin{pmatrix} 2 & 0 & 6 \\ 8 & 7 & 0 \\ 0 & -6 & 0 \end{pmatrix}$$

6) We write the interval: $ds^2 = g_{\mu\nu} dx^\mu dx^\nu$

$= g_{11} dx_1^2 + g_{22} dx_2^2 + g_{33} dx_3^2 + g_{44} dx_4^2$

$+ 2g_{12} dx_1 dx_2 + 2g_{13} dx_1 dx_3 + 2g_{14} dx_1 dx_4$

$+ 2g_{24} dx_2 dx_4 + 2g_{34} dx_3 dx_4 + 2g_{41} dx_4 dx_1$

$+ 2g_{42} dx_4 dx_2 + 2g_{43} dx_4 dx_3$

10) The curvature $\kappa = |dT/ds|$ and we have:

$dT/ds = (dT/dt)/(ds/dt)$

where: $ds/dt = 13$ (from text)

and:

$dT/dt = 1/13 [-24 \sin 2t\mathbf{i} - 24 \cos 2t\mathbf{j}]$

$dT/ds = 1/13 [-24 \sin 2t\mathbf{i} - 24 \cos 2t\mathbf{j}]/ 13$

$dT/ds = [-24 \sin 2t\mathbf{i} - 24 \cos 2t\mathbf{j}]/ 169$

Therefore: $\kappa = |dT/ds| = 24/169$

Since $ds/dt = 13$, the length over the specified interval is:

$$L = \int_0^\pi ds = \int_0^\pi 13 \, dt$$

$s = 13\,t\,]_0{}^\pi = 13\,\pi$

11) We perform the straightforward integration:

$$L := \int_{-\pi}^{\pi} \sqrt{r(\theta)^2 + \left(\frac{d}{d\theta}r(\theta)\right)^2} \, d\theta$$

$L = 8.764$

L= 8.764 units

12) $(ds/dt)^2 = (dx^1/dt)^2 + (x^1)^2 \, (dx^2/dt)^2$

$+ (x^1 \sin x^2)^2 (dx^3/dt)^2$

And:

$(dx^1/dt)^2 = 1$

$(dx^2/dt)^2 = [\,-1/\,t^2 \,/\,\sqrt{\{1 - (1/\,t^2\,)\}} = 1/\,t^2\,(t^2 - 1)$

$(dx^3/dt)^2 = 2t/\sqrt{2}(t^2 - 1) = t^2/(t^2 - 1)$

Whence:

$(ds/dt)^2 =$

$1 + t^2 \bullet 1/t^2 (t^2 - 1) + (t \bullet 1/t)^2 \bullet t^2/(t^2 - 1)$

$= 2t^2/(t^2 - 1)$

Then the length of the arc is:

$L = \int_1^2 \sqrt{2}\, t/(t^2 - 1)^{1/2}\, dt =$

$\sqrt{2(t^2 - 1)}\,]_1^2 = \sqrt{6}$

14) Given the metric: $1, x^1, x^2$

We have:

$R \text{ (Ricci)} = g^{11}(R_{11}) + g^{22}(R_{22}) + g^{33}(R_{33})$

Where: $R_{11} = g^{22} R_{2112}$

$R_{22} = g^{11} R_{1221} + g^{33} R_{3223}$

$R_{33} = g^{22} R_{2332}$

Then: $g_{11} = 1, \quad g_{22} = 1/x^1, \quad g_{33} = 1/x^2$

$R_{1221} = -1/x^1$

$R_{2332} = -1/x^2$

And: $R_{3223} = -1/x^2$

Further: $R_{2112} = -1/x^1$

And: $R_{2332} = -1/x^2$

So:

$R_{11} = g^{22} R_{2112} = 1/x^1 (-1/x^1)$

$R_{11} = -2/(x^1)^2$

$R_{22} = g^{11} R_{1221} + g^{33} R_{3223}$

$= (-1/x^1) + (1/x^2)(-1/x^2)$

$R_{22} = -1/x^1 - 1/(x^2)^2$

$R_{33} = g^{22} R_{2332} = (1/x^1)(-1/x^2)$

$R_{33} = -1/x^1 x^2$

Then: $g^{11}(R_{11}) + g^{22}(R_{22}) + g^{33}(R_{33}) =$

(1) $[-1/(x^1)^2] + 1/x^1 [-1/x^1 - 1/(x^2)^2]$

$+ (1/x^2)[-1/x^1 x^2] =$

$-2/(x^1)^2 - 1/x^1(x^2)^2 - 1/[x^2(x^1 x^2)]$

15) By Stoke's theorem (Mathematical supplement)

$$\int F \bullet dS = \int \nabla \times F \bullet dS$$

$F = -(a_x x\mathbf{i} + a_y y\mathbf{j} + a_z j\mathbf{k})$

Let: $x = \mathbf{i} \cdot \mathbf{r}$ $y = \mathbf{j} \cdot \mathbf{r}$ $z = \mathbf{k} \cdot \mathbf{r}$

Then: substitute and write in eqn. for $\mathbf{F}$:

$$\mathbf{F} = -(a_x \mathbf{i}(\mathbf{i} \cdot \mathbf{r}) + a_y \mathbf{j}(\mathbf{j} \cdot \mathbf{r}) + a_z \mathbf{k}(\mathbf{k} \cdot \mathbf{r}))$$

$$\mathbf{F} = -(a_x \mathbf{i}\mathbf{i} + a_y \mathbf{j}\mathbf{j} + a_z \mathbf{k}\mathbf{k}) \cdot \mathbf{r}$$

$$= -\Phi \cdot \mathbf{r}$$

So: $\int \mathbf{F} \cdot dS = \int -\Phi \cdot \mathbf{r} \cdot dS$

$$\int \nabla \times \mathbf{F} \cdot dS = -\iint \nabla \times \Phi \cdot \mathbf{r} \, dS$$

$$1/r \int -\Phi \cdot \mathbf{r} \cdot dS = -1/r \iint \nabla \times \Phi \cdot \mathbf{r} \, dS$$

Or: $\int \Phi \cdot dS = \iint \nabla \times \Phi \cdot dS$

22) a) The density is expressed:

$$\rho(t) = 3/ 8\pi G t^2$$

Where $G = 6.6726 \times 10^{-11} \, m^3 s^{-2} kg^{-1}$

$t = 0.01 \, s$

$$\rho(t) = 3/ 8\pi [6.6726 \times 10^{-11} \, m^3 s^{-2} kg^{-1}](0.01s)^2$$

$$\rho(t) = 1.78 \times 10^{13} \, kg/m^3$$

b) The result is unaffected because $\rho(t)$ does not depend on H, the Hubble constant.

23) We have: $\alpha = \int_{-\pi/2}^{\pi/2} k M/ r^2 \cos \theta \, ds$

The diagram for reference is:

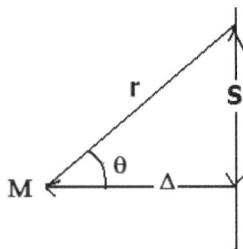

Rewrite: $\alpha = (k M/ r^2) \, 1/c^2 \int_{-\pi/2}^{\pi/2} \cos \theta \, ds$

$\alpha = (k M/ r^2) \, 1/c^2 \int_{-\pi/2}^{\pi/2} \cos \theta \, ds =$

$(k M/ r^2) \, 1/c^2 [\sin \theta]_{-\pi/2}^{\pi/2} =$

$(k M/ r^2) \, 1/c^2 [\sin \pi/2 - \sin (-\pi/2)]$

$= 2 k M/ c^2 r^2$

So: $\alpha = 2k M/ c^2 \Delta = 2GM/ c^2 \Delta$

Note: The triangle simply shows the relationships of distances r, s, and Δ with respect to the center of the body of mass M. Distance s is along the path traveled by the light ray whose bending is given by angle α, and simply shows the ray **at a right angle** to radial

266

distance Δ. It should also be understood that the formula also calls for a *negative version* of the triangle, defining an angle θ that theoretically varies from $-90°$ to $0°$ to $+90°$.

The final version or result actually only makes sense by a change of variable in the integral from s to θ, thereby obtaining;

$$\alpha = (G M / \Delta)\ 1/c^2 \int_{-\pi/2}^{\pi/2} \cos\theta\ d\theta$$

$$= 2GM/ c^2\ \Delta$$

Where we recognize: $\Phi = G M / \Delta$

As the *gravitational potential*. Then we can actually rewrite Einstein's equation as:

$$\alpha = 2\Phi/ c^2$$

Chapter VIII Selected Solutions:

1) From the quantities given:

$$dt/ t \approx GM(1/r_1 - 1/r_2) \approx g(dr)/ c^2$$

where G is the Newtonian gravitational constant, M is the Earth's mass, and g is the acceleration of gravity (g $= 980$ cm/ sec^2 in cgs) and c $= 3 \times 10^{10}$ cm/sec.

From the hypothetical data the box deflection ($r_2 - r_1$)was 0.001 mm = 0.0001cm, then:

$$dt/t \sim (980\ cm/s^2)(10^{-4}\ cm)/ (3 \times 10^{10}\ cm/sec\ )^2$$

$dt/t \approx 10^{-22}$

and for an interval say $t = 0.01$ sec, $dt =$

$(10^{-22})(0.01 \text{ sec}) = 10^{-24}$ sec

The observation would actually generate a time uncertainty of 10^{-24} sec- and hence an uncertainty ΔE in the energy of the photon.,

The mass uncertainty is $\Delta m = \Delta E/ c^2$

$\Delta E\, \Delta t \geq h/ 2\pi$ so $\Delta E \approx 1.054 \times 10^{-34}$ J-s/ 10^{-24} s

$\Delta E \approx 1.054 \times 10^{-10}$ J

Therefore: $\Delta m = \Delta E/ c^2 \approx 1.054 \times 10^{-10}$ J/ 3×10^8 ms^{-1}

$\Delta m \approx 3.5 \times \times 10^{-17}$ kg

The uncertainty in weight is:

$\Delta W \approx \Delta m\ g \approx 3.5 \times \times 10^{-17}$ kg (9.81 N/ kg)

Or: $\Delta W \approx 1.2 \times 10^{-17}$ N

4) According to the authors of **Gravitation and Spacetime** (p. 7) the *upper limit* on the mass of the photon is 10^{-59} g. Assume this to also be the rest mass in the photon's (inertial) rest frame. Find its clock

frequency according to the de Broglie- Bohm-Hiley pilot wave concept.

First find its deBroglie wavelength if it's traveling at

$v = c.$

Then the clock frequency would be:

$\omega_0 = m_0 \, c^2 / \, \hbar =$

$[10^{-59} \text{ kg}] [3 \times 10^{8} \text{ ms}^{-1}] / (1.054 \times 10^{-34} \text{ Js}]$

$= 2.84 \times 10^{-17} \text{ /s}$

De Broglie wavelength: $\lambda_D = h/p = h/ \, m \, c$

$= \quad 6.62 \times 10^{-34} \text{ Js} / [10^{-59} \text{ kg}] [3 \times 10^{8} \text{ ms}^{-1}]$

$= \quad 2.20 \times \times 10^{17} \text{ m}$

5) We know

$\Delta k = \pi \, / \, x = \pi \, / \, (0.5 \text{ nm}) = 2\pi \text{ nm}$, but $2\pi \text{ (nm)} > 2$ nm.

The maximum of the wave packet is approximated closely by the square of the amplitude:

$[E_z]^2 = \quad 4 \sin^2 2\pi \, (1 - 0.5) \, / \, (1 - 0.5) =$

$4 \sin^2 2\pi$

But: $\sin 2\pi = 0$ so $[E_z]^2 = 0$

Physical interpretation: The dimension of Δk relative to: $\Delta x = (x - x_0)$

Implies no wave packet can exist.

6) To test Bohmian quantum mechanics on a computer, the uncertainty in one input turns out to be: $\Delta t = 10^{-39}$ s and in the other, $\Delta \phi_k = 10^{-51}$ m. From this data, we can find the quantity b. We first compose a form of the Uncertainty principle and obtain the product ab. Then:

$$(\delta\ \Delta\phi_k)^2 = b\ (\Delta t)$$

$$(\delta\ \Delta\phi_k) = b^{1/2}\ (\Delta t)^{1/2}$$

$$b = (\delta\ \Delta\phi_k)^2\ /\ (\Delta t) = (10^{-51}\ m)^2 / (10^{-39}\ s) =$$

$$(10^{-63}\ m)^2\ /\ s$$

Note: a is a constant of proportionality so let a = 1

Bohm notes that π_k also fluctuates at random over the given range so:

$$\delta\,\pi_k = a\,b^{1/2} /\ (\Delta t)^{1/2} =$$

$$(10^{-51}\ m)^{1/2}\ /\ (10^{-39}\ s)^{1/2} = 10^{-6}\ m^{1/2}\ s^{-1/2}$$

Combining all the preceding results one finally has:

$$\delta\,\pi_k\ (\delta\ \Delta\phi_k) = ab = (10^{-6}\ m^{1/2}\ s^{-1/2})(10^{-51}\ m)$$

$$= 10^{-57}\ m^{3/2}\ s^{-1/2}\quad \text{Analogous to: } \delta p\,\delta q \le \hbar$$

Mathematical Supplement

1. *Basic Algebra*:

$x^a \cdot x^b$ simplifies to $x^{(a+b)}$

$(x \cdot y)^a$ simplifies to $x^a \cdot y^a$

$\left(x^a\right)^b$ simplifies to $x^{(a \cdot b)}$

$\dfrac{1}{x^a}$ simplifies to $x^{(-a)}$

$\dfrac{x^a}{x^b}$ simplifies to $x^{(a-b)}$

$(a+b)^2$ expands to $a^2 + 2 \cdot a \cdot b + b^2$

$a^3 - b^3$
by factoring, yields $(a-b) \cdot \left(a^2 + a \cdot b + b^2\right)$

Solving $a \cdot x^2 + b \cdot x + c = 0$ for x

has solution(s)
$$\left[\begin{array}{c} \dfrac{1}{(2 \cdot a)} \cdot \left(-b + \sqrt{b^2 - 4 \cdot a \cdot c}\right) \\[3mm] \dfrac{1}{(2 \cdot a)} \cdot \left(-b - \sqrt{b^2 - 4 \cdot a \cdot c}\right) \end{array} \right]$$

Properties of Logarithms

$\ln(x \cdot y)$ simplifies to $\ln(x) + \ln(y)$

$\ln\left(\dfrac{x}{y}\right)$ simplifies to $\ln(x) - \ln(y)$

$\ln\left(x^a\right)$ simplifies to $a \cdot \ln(x)$

2. Basic Geometry:

2.1. Parallelogram:

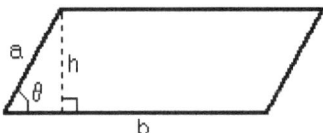

Expressions for Area:

$b \cdot h$ area in terms of b and h

$a \cdot b \cdot \sin(\theta)$ area in terms of a, b, and θ

Perimeter: $2 \cdot a + 2 \cdot b$

2.2 Scalene Triangle:

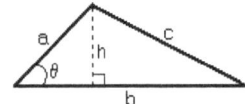

Expressions for Area:

$\dfrac{1}{2} \cdot h \cdot b$ area in terms of h and b

$\sqrt{s \cdot (s - a) \cdot (s - b) \cdot (s - c)}$ Heron's formula, where s is the semiperimeter

$\dfrac{1}{2} \cdot a \cdot b \cdot \sin(\theta)$ area in terms of a, b, and θ

Perimeter: $a + b + c$

2.3 Circle & Sectors

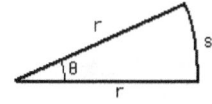

Area: $\pi \cdot r^2$

Area: $\frac{1}{2} \cdot r^2 \cdot \theta$

Perimeter: $2 \cdot \pi \cdot r$ Arc length: $r \cdot \theta$

2.4
Right triangle relationships

$\sin(\theta) = \dfrac{\text{opp}}{\text{hyp}}$

$\cos(\theta) = \dfrac{\text{adj}}{\text{hyp}}$

$\tan(\theta) = \dfrac{\text{opp}}{\text{adj}}$

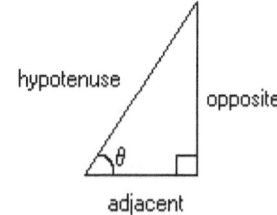

2.5. <u>Trigonometric functions by quadrant</u>:

	$-\theta$	$\frac{\pi}{2} \pm \theta$	$\pi \pm \theta$	$\frac{3\pi}{2} \pm \theta$	$2k\pi \pm \theta$
sin	$-\sin\theta$	$\cos\theta$	$\mp\sin\theta$	$-\cos\theta$	$\pm\sin\theta$
cos	$\cos\theta$	$\mp\sin\theta$	$-\cos\theta$	$\pm\sin\theta$	$+\cos\theta$
tan	$-\tan\theta$	$\mp\cot\theta$	$\pm\tan\theta$	$\mp\cot\theta$	$\pm\tan\theta$
csc	$-\csc\theta$	$+\sec\theta$	$\mp\csc\theta$	$-\sec\theta$	$\pm\csc\theta$
sec	$\sec\theta$	$\mp\csc\theta$	$-\sec\theta$	$\pm\csc\theta$	$+\sec\theta$
cot	$-\cot\theta$	$\mp\tan\theta$	$\pm\cot\theta$	$\mp\tan\theta$	$\pm\cot\theta$

Where: $0 \le \theta \le \pi/2$

273

2.6. Trigonometric Values for Defined Angles:

	0 $0°$	$\pi/12$ $15°$	$\pi/6$ $30°$	$\pi/4$ $45°$	$\pi/3$ $60°$
sin	0	$\frac{\sqrt{2}}{4}(\sqrt{3}-1)$	$1/2$	$\sqrt{2}/2$	$\sqrt{3}/2$
cos	1	$\frac{\sqrt{2}}{4}(\sqrt{3}+1)$	$\sqrt{3}/2$	$\sqrt{2}/2$	$1/2$
tan	0	$2-\sqrt{3}$	$\sqrt{3}/3$	1	$\sqrt{3}$
csc	∞	$\sqrt{2}(\sqrt{3}+1)$	2	$\sqrt{2}$	$2\sqrt{3}/3$
sec	1	$\sqrt{2}(\sqrt{3}-1)$	$2\sqrt{3}/3$	$\sqrt{2}$	2
cot	∞	$2+\sqrt{3}$	$\sqrt{3}$	1	$\sqrt{3}/3$

	$5\pi/12$ $75°$	$\pi/2$ $90°$	$7\pi/12$ $105°$	$2\pi/3$ $120°$
sin	$\frac{\sqrt{2}}{4}(\sqrt{3}+1)$	1	$\frac{\sqrt{2}}{4}(\sqrt{3}+1)$	$\sqrt{3}/2$
cos	$\frac{\sqrt{2}}{4}(\sqrt{3}-1)$	0	$\frac{-\sqrt{2}}{4}(\sqrt{3}-1)$	$-1/2$
tan	$2+\sqrt{3}$	∞	$-(2+\sqrt{3})$	$-\sqrt{3}$
csc	$\sqrt{2}(\sqrt{3}-1)$	1	$\sqrt{2}(\sqrt{3}-1)$	$2\sqrt{3}/3$
sec	$\sqrt{2}(\sqrt{3}+1)$	∞	$-\sqrt{2}(\sqrt{3}+1)$	-2
cot	$2-\sqrt{3}$	0	$-(2-\sqrt{3})$	$-\sqrt{3}/3$

2.6 continued:

	$3\pi/4$ $135°$	$5\pi/6$ $150°$	$11\pi/12$ $165°$	π $180°$
sin	$\sqrt{2}/2$	$1/2$	$\dfrac{\sqrt{2}}{4}(\sqrt{3}-1)$	0
cos	$-\sqrt{2}/2$	$-\sqrt{3}/2$	$\dfrac{-\sqrt{2}}{4}(\sqrt{3}+1)$	-1
tan	-1	$-\sqrt{3}/3$	$-(2-\sqrt{3})$	0
csc	$\sqrt{2}$	2	$\sqrt{2}(\sqrt{3}+1)$	∞
sec	$-\sqrt{2}$	$-2\sqrt{3}/3$	$-\sqrt{2}(\sqrt{3}-1)$	-1
cot	-1	$-\sqrt{3}$	$-(2+\sqrt{3})$	∞

Graphs of Trigonometric functions:

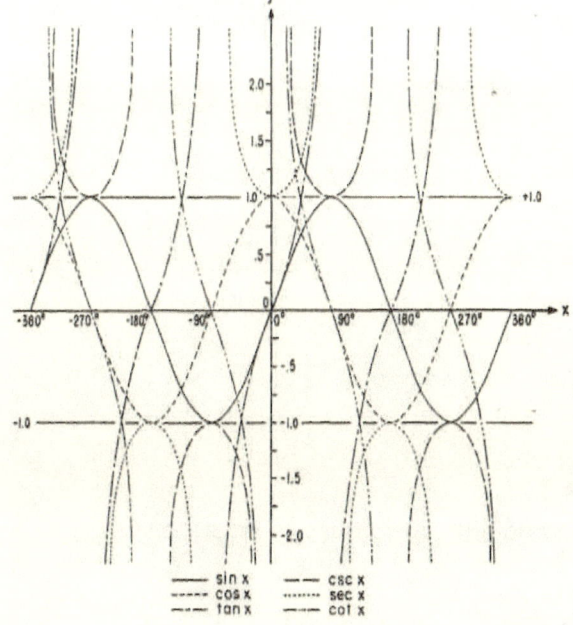

275

Right triangle relationships

$$\sin(\theta) = \frac{opp}{hyp}$$

$$\cos(\theta) = \frac{adj}{hyp}$$

$$\tan(\theta) = \frac{opp}{adj}$$

hypotenuse

opposite

θ

adjacent

$\tan(x)$ simplifies to $\dfrac{\sin(x)}{\cos(x)}$

$\cot(x)$ simplifies to $\dfrac{\cos(x)}{\sin(x)}$

$\sec(x)$ simplifies to $\dfrac{1}{\cos(x)}$

$\csc(x)$ simplifies to $\dfrac{1}{\sin(x)}$

$\sin(\theta)^2 + \cos(\theta)^2$ simplifies to 1

$\tan(x)^2 + 1$ simplifies to $\dfrac{1}{\cos(x)^2}$

$\sin(2 \cdot x)$ expands to $2 \cdot \sin(x) \cdot \cos(x)$

$\cos(x + y)$

expands to $\cos(x) \cdot \cos(y) - \sin(x) \cdot \sin(y)$

$\dfrac{1}{2} \cdot (1 - \cos(2 \cdot t))$ expands to $1 - \cos(t)^2$

Basic Trigonometric Equalities:

$\sin (x \pm y) = \sin x \cos y \pm \cos x \sin y$
$\cos (x \pm y) = \cos x \cos y \pm \sin x \sin y$

$\sin^2 x = \frac{1}{2}(1 - 2\cos x)$

2.7. Basic Differentiation formulae:

$\ln(\lvert x \rvert)$	by differentiation, yields	$\dfrac{1}{x}$
$\sin(x)$	by differentiation, yields	$\cos(x)$
$\text{atan}(x)$	by differentiation, yields	$\dfrac{1}{\left(1 + x^2\right)}$
$\cosh(x)$	by differentiation, yields	$\sinh(x)$

Additional Differentiation formulae:

$$\frac{d}{dz} e^z = e^z$$

$$\frac{d^n}{dz^n} e^{az} = a^n e^{az}$$

$$\frac{d}{dz} a^z = a^z \ln a$$

$$\frac{d}{dz} z^a = a z^{a-1}$$

$$\frac{d}{dz} z^z = (1 + \ln z) z^z$$

2.6 Integration formulae:

Basic:

$\cos(u)$ by integration, yields $\sin(u)$

$\sec(u) \cdot \tan(u)$ by integration, yields $\dfrac{1}{\cos(u)}$

b^u by integration, yields $\dfrac{1}{\ln(b)} \cdot b^u$

More complex:

$$\int \frac{dz}{z} = \ln z$$

$$\int \ln z \, dz = z \ln z - z$$

$$\int e^{az} dz = e^{az}/a$$

$$\int \sin z \, dz = -\cos z$$

$$\int \cos z \, dz = \sin z$$

$$\int \tan z \, dz = -\ln \cos z = \ln \sec z$$

Stokes' Theorem:

$$\int F \bullet dS \;=\; \int \nabla \times F \bullet dS$$

3.1 Taylor Series:

Let f be a function defined on the closed interval [a, b] and f has n continuous derivatives on [a, b] then f^{n+1} (x) exists for all x in [a, b]. And:

$f(b) = f(a) + f\,'(a) \cdot (b - a) + f\,''\,(a)/\,2!\,(b - a)^2 +$

$....+\ f^{\underline{n}}\,(a)\,/\,n!\,(b - a)^n + f^{n+1}\,(c)/\,(n + 1)!\,[(b - a)^{n+1}]$

Where c lies between a and b.

Example: Integrate: $\int^{b}_{a} f(x)\ dx$

Using a *Taylor series* method and justify it.

Take $f(x) = \sum^{\infty}_{n=0}\ \ f^{\underline{n}}\,(a)\,/\,n!\,(x - a)^{n}$

Then: $\int^{b}_{a}\ \ f(x)\ dx =$

$\int^{b}_{a}\ \ f(a)\ dx + \int^{b}_{a}\ f\,'(a) \cdot (x - a)\ \ dx$

$+ \int^{b}_{a}\ \ f\,''\,(a)/\,2!\,(x - a)^2$

$\int^{b}_{a}\ f(x)\ dx = [f\,(a) \cdot (b - a)\ +\ f\,'(a)/\,2!\,(x - a)^2\ |^{b}_{a}$

$+ \int^{b}_{a}\ \ f\,''\,(c)/\,2!\,(x - a)^2$

The addition of the two terms in the bracket, with b, a limits, will yield a least upper bound to the value of the integral. Given a function f, assume it is infinitely differentiable at some point a. Then we can write:

$$f(x) = \sum\nolimits_{n=0}^{\infty} \; f^{\underline{n}}(a) \,/\, n! \,(x-a)^n$$

3.2. VECTOR IDENTITIES

Notation: f, g, are scalars; A, B, etc., are vectors; T is a tensor; I is the unit dyad.

(1) $A \cdot B \times C = A \times B \cdot C = B \cdot C \times A = B \times C \cdot A = C \cdot A \times B = C \times A \cdot B$

(2) $A \times (B \times C) = (C \times B) \times A = (A \cdot C)B - (A \cdot B)C$

(3) $A \times (B \times C) + B \times (C \times A) + C \times (A \times B) = 0$

(4) $(A \times B) \cdot (C \times D) = (A \cdot C)(B \cdot D) - (A \cdot D)(B \cdot C)$

(5) $(A \times B) \times (C \times D) = (A \times B \cdot D)C - (A \times B \cdot C)D$

(6) $\nabla(fg) = \nabla(gf) = f\nabla g + g\nabla f$

(7) $\nabla \cdot (fA) = f\nabla \cdot A + A \cdot \nabla f$

(8) $\nabla \times (fA) = f\nabla \times A + \nabla f \times A$

(9) $\nabla \cdot (A \times B) = B \cdot \nabla \times A - A \cdot \nabla \times B$

(10) $\nabla \times (A \times B) = A(\nabla \cdot B) - B(\nabla \cdot A) + (B \cdot \nabla)A - (A \cdot \nabla)B$

(11) $A \times (\nabla \times B) = (\nabla B) \cdot A - (A \cdot \nabla)B$

(12) $\nabla(A \cdot B) = A \times (\nabla \times B) + B \times (\nabla \times A) + (A \cdot \nabla)B + (B \cdot \nabla)A$

3) $\nabla^2 f = \nabla \cdot \nabla f$

(14) $\nabla^2 A = \nabla(\nabla \cdot A) - \nabla \times \nabla \times A$

(15) $\nabla \times \nabla f = 0$

(16) $\nabla \cdot \nabla \times A = 0$

Basic Vector Notation:

Consider the vector diagram below:

280

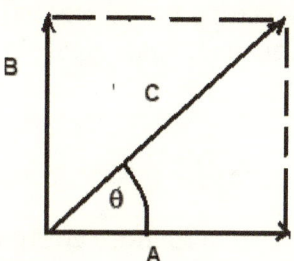

Here we may write: $\mathbf{C} = \mathbf{A} + \mathbf{B}$

The magnitude of the resultant C is:

$$|\mathbf{C}| = \sqrt{(A^2 + B^2)}$$

With no loss of generality we can change the vector components and resultant thus:

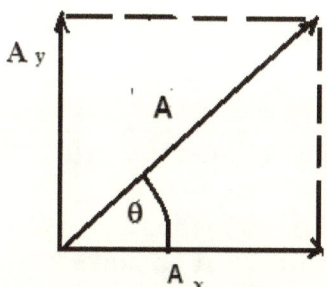

Here, the resultant is $\mathbf{A}$ and the components are: $\mathbf{A}_x$ and $\mathbf{A}_y$ such that: $\mathbf{A} = \mathbf{A}_x + \mathbf{A}_y$ where:

$$\mathbf{A}_x = A_x \mathbf{i} \quad \text{and } \mathbf{A}_y = A_y \mathbf{j}$$

Where $\mathbf{i}$ and $\mathbf{j}$ are unit vectors, and:

$$A_x = A\cos\theta, \quad A_y = A\sin\theta, \quad \theta = \tan^{-1}(A_y / A_x)$$

A_x and A_y are *scalar* quantities called the *components* of the vector **A**. Again, the magnitude of **A** can be written:

$$|A| = \sqrt{(A_x^2 + A_y^2)}$$

It is also possible to have more complex vector configurations which require more analysis, such as shown in the diagram below:

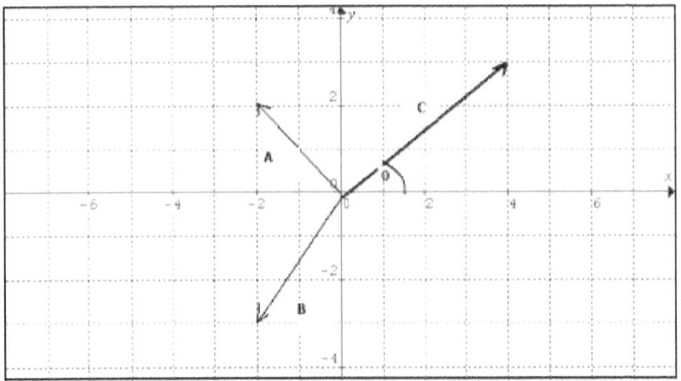

Fig. 3.1 Diagram showing multiple resultants

In Fig. 3.1 the task is to identify the components for each of the resultants, **A, B** and **C**. Given the units are provided to scale, e.g. the resultant A has two equal components of 2 units each, this is possible.

Then for resultant **A**:

$$A = A_x \mathbf{i} + A_y \mathbf{j}$$

$$A_x = -2, \quad A_y = 2,$$

$$|A| = \sqrt{(A_x^2 + A_y^2)} = \sqrt{(-2)^2 + (2)^2}$$

$|\mathbf{A}| = \sqrt{(8)} = 2\sqrt{2}$

$\theta = \tan^{-1}(A_y / A_x) = \tan^{-1}(2 / -2) = \tan^{-1}(-1)$

$\theta = -45$ degrees

Similarly:

$\mathbf{B} = B_x \mathbf{i} + B_y \mathbf{j}$

$B_x = -2, \quad B_y = -3,$

$|\mathbf{B}| = \sqrt{(B_x{}^2 + B_y{}^2)} = \sqrt{(-2)^2 + (-3)^2}$

$|\mathbf{B}| = \sqrt{(13)}$

$\theta = \tan^{-1}(B_y / B_x) = \tan^{-1}(-3 / -2) = \tan^{-1}(3/2)$

$\theta = 56.3$ degrees

Finally, we write for resultant $\mathbf{C}$:

$|\mathbf{C}| = \sqrt{(C_x{}^2 + C_y{}^2)} = \sqrt{(4)^2 + (3)^2}$

$|\mathbf{C}| = \sqrt{(25)} = 5$ units

$\theta = \tan^{-1}(C_y / C_x) = \tan^{-1}(3 / 4) = \tan^{-1}(3/4)$

$\theta = 36.8$ degrees

More generally, we can have a situation (e.g. in 3 dimensions) with several vectors to be added:

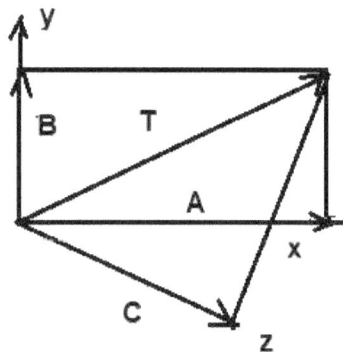

So in this case: $\mathbf{T} = \mathbf{A} + \mathbf{B} + \mathbf{C}$

Whence:

$$T_x = A_x + B_x + C_x$$

$$T_y = A_y + B_y + C_y$$

$$T_z = A_z + B_z + C_z$$

Then once the components of **T** have been found the magnitude of the vector may be obtained:

$$\| \mathbf{T} \| = \sqrt{(T_x^2 + T_y^2 + T_z^2)}$$

By definition the components of such a vector are numbers which multiply the unit vectors . So if the components of **T** are T_x, T_y, and T_z then:

$$\mathbf{T} = T_x \mathbf{i} + T_y \mathbf{j} + T_z \mathbf{k}$$

We can also have the situation for which $\mathbf{T} = \mathbf{A} + \mathbf{B},$ but for which the two vectors, A and B, are not necessarily in the same plane. Hence each vector can be described in 3 dimensions using unit vectors. Then one can write:

$$\mathbf{T} = (A_x\mathbf{i} + A_y\mathbf{j} + A_z\mathbf{k}) + (B_x\mathbf{i} + B_y\mathbf{j} + B_z\mathbf{k})$$

And since vector addition is commutative, the preceding can also be written:

$$\mathbf{T} = (A_x + B_x)\mathbf{i} + (A_y + B_y)\mathbf{j} + (A_z + B_z)\mathbf{k}$$

We note the length of any vector in 3 dimensions, say referenced along the diagonal of a three- dimensional box, say:

$$\mathbf{T} = a\mathbf{i} + b\mathbf{j} + c\mathbf{k}$$

Is easily determined by applying the Pythagorean theorem twice, once to the diagonal of a face the rectangular box, then to the overall diagonal, e.g.

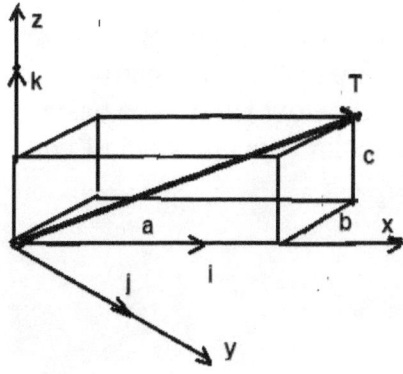

Fig. 3.2.

In summary:

$$\| a\,\mathbf{i} + b\,\mathbf{j} + c\,\mathbf{k} \| = \sqrt{(a^2 + b^2 + c^2)}$$

Direction cosines:

If the diagonal vector $\mathbf{T} = a\,\mathbf{i} + b\,\mathbf{j} + c\,\mathbf{k}$ makes angles α, β and γ ,respectively, i.e. with the x-, y- and z- axes, then: $\cos\alpha$, $\cos\beta$ and $\cos\gamma$ arc called the direction cosines, where:

$$\cos\alpha = a / \sqrt{(a^2 + b^2 + c^2)}$$

$$\cos\beta = b / \sqrt{(a^2 + b^2 + c^2)}$$

$$\cos\gamma = b / \sqrt{(a^2 + b^2 + c^2)}$$

It can also be shown that:

$$\cos^2\alpha + \cos^2\beta + \cos^2\gamma = 1$$

Dot Product:

The scalar product of two vectors **A** and **B** is also called the *dot product* because of the dot symbol used to denote it. In the diagram below I show the basic geometry relevant to the scalar (dot) product of two vectors, say **A** **·B.**

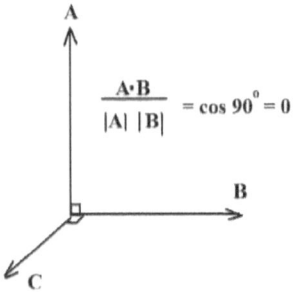

$$\frac{A \cdot B}{|A|\,|B|} = \cos 90^{0} = 0$$

Thus:

$\mathbf{A} \cdot \mathbf{B} = |\mathbf{A}||\mathbf{B}| \cos \Theta$,, hence also:

$\cos(\Theta) = (\mathbf{A} \cdot \mathbf{B})/ |\mathbf{A}||\mathbf{B}|$

If the two vectors are at a right angle to each other then Θ) = 90 degrees and:

$\cos(\Theta) = (\mathbf{A} \cdot \mathbf{B})/ |\mathbf{A}||\mathbf{B}| = 0$

Since vector multiplication is commutative, we also have:

$\mathbf{A} \cdot \mathbf{B} = \mathbf{B} \cdot \mathbf{A}$

Cross Product:

The vector cross product $\mathbf{A} \times \mathbf{B}$ is illustrated below:

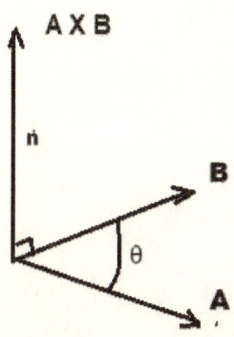

Here, we let the angle subtended between **A** and **B** be Θ with: $0 \leq \Theta \leq \pi$. Then unless **A** and **B** are parallel, they now determine a plane. Let **n** be a unit vector perpendicular to the plane and pointing in the direction a right-handed thread screw would advance when its head is rotated from **A** to **B** through the angle Θ. The vector product or cross product is then:

A X B   =   **n** |**A** ||**B**| sin Θ

It is clear that if A and B are parallel (Θ = 0) so:

A X B   =   0

Also, if **A** and **B** are reversed in the depiction above, then it follows the unit vector **n** is replaced by − **n** and hence:

B X A   = - **A X B**

Let us say the vectors are respectively represented by:

$A = a_x \mathbf{i} + a_y \mathbf{j} + a_z \mathbf{k}$

$B = b_x \mathbf{i} + b_y \mathbf{j} + b_z \mathbf{k}$

Then the cross product can be conveniently written as a 3rd order determinant:

$$A \times B = \begin{bmatrix} \mathbf{i} & \mathbf{j} & \mathbf{k} \\ a_x & a_y & a_z \\ b_x & b_y & b_z \end{bmatrix}$$

Triple Scalar Product:

We refer to the product: (**A X B**) · **C** as *the triple scalar produc*t, given that it is derived from vector dimensions in a solid geometry setting. On inspection of the diagram below, of a paralleilepiped, we see the vector $\mathbf{N} = \mathbf{A} \times \mathbf{B}$ is normal to the base (determined by the vectors $\mathbf{A}$, $\mathbf{B}$) so equals to the area of that base. Thus:

$$(\mathbf{A} \times \mathbf{B}) \cdot \mathbf{C} \quad = \quad |\mathbf{N}||\mathbf{C}| \cos \Theta$$

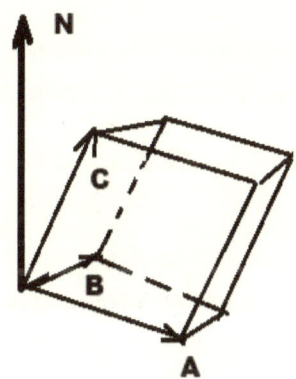

Thus: $\mathbf{N} = \mathbf{A} \times \mathbf{B}$ = area of the base

And: $|\mathbf{C}| \cos \Theta$ = altitude of the box

Given $\mathbf{A}$ is defined:

$$\mathbf{A} = a_1\mathbf{i} + a_2\mathbf{j} + a_3\mathbf{k}$$

And:

$$\mathbf{B} = b_1\mathbf{i} + b_2\mathbf{j} + b_3\mathbf{k}$$

And $\mathbf{C}$ similarly, with:

$$(\mathbf{A} \times \mathbf{B}) \cdot \mathbf{C} = \mathbf{A} \cdot \mathbf{B} \times \mathbf{C}$$

The triple scalar product can be expressed:

$$(\mathbf{A} \times \mathbf{B}) \cdot \mathbf{C} \quad = \begin{bmatrix} a_1 & a_2 & a_3 \\ b_1 & b_2 & b_3 \\ c_1 & c_2 & c_3 \end{bmatrix}$$

Triple Vector Product:

Represent the product of three vectors **A, B** and **C** with two equations that are companions of each other:

$$(\mathbf{A} \times \mathbf{B}) \times \mathbf{C} \; = \; (\mathbf{A} \cdot \mathbf{C})\,\mathbf{B} \; - \; (\mathbf{B} \cdot \mathbf{C})\,\mathbf{A}$$

And:

$$\mathbf{A} \times (\mathbf{B} \times \mathbf{C}) \; = \; (\mathbf{A} \cdot \mathbf{C})\,\mathbf{B} \; - \; (\mathbf{A} \cdot \mathbf{B})\,\mathbf{C}$$

Matrices and Eigenvalues:

a)**Matrix Operation Basics:**

The simple operation of matrix multiplication such that, given a matrix **A**:

$(a_{11} \; a_{12})$
$(a_{21} \; a_{22})$

and a matrix **B**:

$(b_{11} \; b_{12})$
$(b_{21} \; b_{22})$

then **A** X **B** =

(a11 a12) (b11 b12)
(a21 a22) (b21 b22)

= [{(a11b11) + (a12b21)} --{(a11 b12)+ (a12 b22)}]
  [(a21b11) + (a22b21) } --{((a21 b12) + (a22 b22)}]

For example, let **A** =

(1 2)
(1 2)

and **B** =

(1 3)
(2 2)

then **A x B** =

(5 7)
(5 7)

The reader should be able to work this out using the format shown. We now want to extend this to further operations with matrices, and we will confine attention to 2 x 2 matrices as subset of **R** ^{2}.

Not all matrices multiply commutatively.

For example, with regular numbers it is a given that:
(2 x 3) = (3 X 2) = 6

Thus, in symbolic form: a x b = b x a and we say the multiplication is commutative. But this need not be so with matrices and matrix multiplication.

For example, let: **A** =
(1 2)
(3 -1)

and **B** =
(2 0)
(1 1)

We find: **A** X **B** =

(4 2)
(5 -1)

But: **B** X **A** =

(2 4)
(4 1)

so that: **A X B** ≠ **B X A** and matrix multiplication doesn't give the same result both ways.

Another application is to obtain the transpose of a

matrix and repeat such multiplication. The transpose of a matrix M, call it t^M, is obtained by the following procedure:

Let M =

(m11......m12)
(m21.....m22)

Then t^M is obtained by *switching the elements* such that for *the transposed* matrix:

m11 = m11,

m12 = m21

m21 = m12   *and* : m22 = m22

 Let **A** =

(2...1)
(3...1)

Find: t^**A**:

Using the procedure shown above for the elements, we have t^**A** =

(2.....3)
(1.....1)

Lastly, we come to the trace of a matrix. This is simply the addition of its diagonal elements. Thus, for any matrix M such as denoted above:

$Tr(M) = m_{11} + m_{22}$

The beauty of this is that it can easily be extended for any dimension matrix, say 3 x 3, or 4 x 4 or whatever. You simply add the diagonal elements:

Find the trace of M_1 =

$(-1.....0.......0)$
$(0.......-1.....0)$
$(0.......0.......1)$

We easily see the diagonal elements and thence add them:

$Tr(M_1) = (-1) + (-1) + 1 = -2 + 1 = -1$

In this linear algebra context, let a 3 x 3 matrix **A** =

$(a_1.....0.......0)$
$(0.......a_2.....0)$
$(0.......0......a_n)$

We're first interested in obtaining its characteristic polynomial from:

$P\_A(t) =$

$$
\begin{pmatrix}
t- a_1 & \dots & 0 & \dots & 0 \\
0 & \dots & t - a_2 & \dots & 0 \\
0 & \dots & 0 & \dots & t - a_n
\end{pmatrix}
$$

Or:

$P\_A(t) = (t - a_1)(t - a_2)\,(t - a_n)$

The *eigenvalues* can be obtained via solving for a_1, a_2, a_n, in the equation:

$(t - a_1)(t - a_2)\,(t - a_n) = 0$

Example Problem:
Given the matrix:
A =
$$
\begin{pmatrix}
1 & \dots & i \\
i & \dots & -2
\end{pmatrix}
$$

Find the characteristic polynomial as well as the eigenvalues.

Solution:
We have: $P\_A(t) =$
$$
\begin{pmatrix}
t - 1 & \dots & i \\
i & \dots & t + 2
\end{pmatrix}
$$

Whence: $P\_A(t) = (t - 1)(t + 2) - (i)^2$

$P\_A(t) = t^2 - t + 2t - 2 - (i)^2 = t^2 + t - 2 + 1 = t^2 + t - 1$

Since this is a *quadratic equation*, so we can find the eigenvalues ($E_{1,2}$) using the quadratic formula:

$E_{1,2} = \sqrt{\{-b \pm [b^2 - 4\,ac]\}} / 2a$

Where a, b, c denote the coefficients for the quadratic, with a the numerical coefficient for the exponent 2 term (t^2), b for the exponent 1 term(t) and c the exponent 0 term. Thus: a = 1, b = 1, c = -1

Then: $E_{1,2} = \sqrt{\{-1 +/- [1^2 - 4(-1)]\}} / 2(1)$

$E_{1,2} = \sqrt{\{-1 +/- [5]\}} / 2$

So that: $E_1 = \sqrt{(-1 + [5])} / 2 = 0.618$

$E_2 = \sqrt{(-1 - [5])} / 2 = -1.618$

Differential Equations

Basic Definitions:

There is no gainsaying the fact that differential equations are the core language for most branches of physics. If one can therefore master the techniques for solving these equations, he is well on the way to mastering most branches of physics. The logical place to start is with definitions.

Differential equation: Any equation containing differentials or derivatives, e.g. dx, dy or dy/dx. If only one variable is involved, the derivatives and differentials are total derivatives, and total differentials.

Ordinary differential equation: Any differential equation which contains total differentials and total derivatives.

Examples include all the following:

   a) $dy/dx = x^3 + 4$

     b) $(x^2 + y^2)dx - 2y\,dy = 0$

c) $d^4y/dt^4 - 2d^3y/dt^3 - 7\,d^2y/dt^2 + 20\,dy/dt - 12\,y = 0$

The order of a differential equation is the order of the highest order derivative in its expression.

Based on this definition, we can see that equations (a) and (b) above are *first order* differential equations, and example (c) is a *fourth order* differential equation.

Aside: It is well to point out here that differential equations can be presented in different forms. Thus, examples (a) and (c) above are written using derivatives, while (b) is in terms of differentials, dx, dy. We can also write using another form that has more or less gone out of use:

$D_x u = ux$ e.g. $du/dx = ux$

Also, example (c) can be written:

$y^{iv} - 2y^{iii} - 2y''' - 7^{ii} + 20y^{i} - 12\,y = 0$

Where each (common) Roman numeral exponent is taken to be a derivative, so: $20y^{i} = 20\,dy/dx$.

Yet another form for differential equations can be expressed:

$f(x, y, y', y'') = 0$

- Indicating an equation involving the independent variable x, the dependent variable y and the first and second derivatives of y. Hence, it can be recast as:

$d^2y/dx^2 + dy/dx + y + x = 0$

The *degree* of an ordinary differential equation is the algebraic degree of the highest ordered derivative.

Thus, from the previous cases, examples (a) and (b) are first degree ordinary differential equations, as is:

$D_x u = ux$

While example (c) is a 4th degree ordinary differential equation, of the 4th order

A linear differential equation is one in which the dependent variable and any of its derivatives have no degree higher than the first.

Ordinary, First Order Differential Equations:

Let's begin with the simplest first order equation imaginable, which is also variables separable:

$$dy = x \, dx$$

As with all differential equations, the solution is accomplished via the process called *integration*.

If we integrate both sides, we obtain:

$$y = x^2 / 2 + c$$

where c is some undefined (as yet) constant of integration. We call the above the "general solution" to the differential equation. This general solution is, in fact, a family of parabolas.

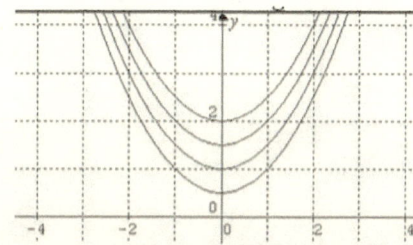

Fig. 1 General solutions for dy = x dx.

If we wanted to obtain the particular solution, we'd have to have *boundary conditions* available. Usually these designate what values x, y are to have at a particular point, and also often the first derivative (y' or dy/dx) at the same point.

Thus, we enter the world of first order differential equations of the first degree.

We consider the equation:

$$x\,dx + y\,dy = 0$$

One might understandably be tempted to write this in terms of the 1st derivative to obtain:

$$dy/dx = -x/y$$

but this serves no useful purpose. It's more productive to simply integrate the equation:

$$x\,dx = -\,y\,dy$$

Thereby obtaining:

$$x^2/2 \;=\; -\,y^2/2 + c \quad \text{or:}$$

$$x^2 + y^2 \;=\; r^2 \;=\; 2c$$

Where $r^2 = 2c$ is the *constant of integration.*

The solution then is the equation of a circle with the center at the origin. For chosen values of r tthis will give a continuous unque curve in the (x,y) –plane and one can show one such solution is:

$$y = \sqrt{(r^2 - x^2)}$$

From this exercise we can glean that: A solution of an ordinary differential equation in two variables is a functional relation between the two variables which satisfies the differential equation. In other words, the primary difference, say to solving a normal algebraic equation, e.g. $x^2 + 2 + 4$

Is that instead of numbers one now obtains a functional relation.

Another basic example is from kinematics: We have the basic differential equation for rate of displacement:

$dx/dt = v$ So we integrate:

$$x = \int (dx/dt)\, dt = \tfrac{1}{2} a t^2 + C$$

Let position $x_o = 0$, e.g. at time $t = 0$

Then on integration:

$$x = \int (v)\, dt = \int (at)\, dt = \tfrac{1}{2} a t^2 + C$$

But: $C = x_o$ so the general solution becomes

$$x = \tfrac{1}{2} a t^2 + x_o$$

And the *particular* solution is (since $x_0 = 0$)

$$x = \tfrac{1}{2} a t^2$$

(Where $(x - x_0)$ is just the displacement).

Now we examine a more complex example: Find the general solution and the particular curve passing through the point (0,0) of the differential equation:

$$[\exp(x) \cos(y)] \, dx + (1 + \exp(x)) \sin(y)] \, dy = 0$$

The first rule is simplify, which means *separating variables* . So we obtain:

$$[\exp(x)/ (1 + \exp(x))] \, dx + [\sin(y)/ \cos(y)] \, dy$$

$$= [\exp(x)/ (1 + \exp(x))] \, dx + \tan(y) dy = 0$$

We then integrate this to obtain:

$$\ln(1 + e^x) - \ln \cos(y) = \ln c$$

Or:

$$\ln(1 + e^x) = \ln c + \ln \cos(y)$$

where again, c is the constant of integration. We can easily simplify the above (using well known principles of natural logs) to get:

$1 + e^x = c \cos(y)$

And this is *the general solution.*

To get the particular solution we need to substitute the ordered pair values for (0,0) into the general solution, whence:

$1 + e^0 = c(\cos(0))$

so that: $1 + 1 = c$

and $c = 2$

Then we get: $1 + e^x = 2\cos(y)$

Integrating factors to Solve DEs:

The differential equation:

$M(x,y)\,dx + N(x,y)\,dy = 0$

Can always be transformed into an exact DE by multiplying it by some suitable factor, call it r(x,y). This makes the DE exact and is called an *"integrating factor".* *Usually* an appropriate r(x,y) can be found on inspection of the DE and visualizing how it might be most directly simplified, say if both sides were multiplied through by some expression.

Example Problem: Find an integrating factor for the DE:

$xdy + ydx = x^2y^2 \, dx$ and solve.

We can rewrite as:

$xdy = [x^2y^2 - y] \, dx$

With $M = x$ and $N = [x^2y^2 - y]$, we easily see on inspection that it's *not exact*.

Leaving the equation as is, one can see that multiplying both sides by $r(x,y) = 1/(x^2y^2)$, will work wonders.

The right hand side simply becomes dx. E.g.

$r(x,y)[xdy + ydx] = dx = 1/(x^2y^2) \, [\, xdy + ydx]$

On close inspection the savvy reader will quickly spot that the more complicated side of the DE is easily reducible *via the exact differential form:*

$1/(x^2y^2) \, [\, xdy + ydx] = d(- 1/xy)$

Then, on inspection, our DE quickly reduces to:

$d(-1/xy) = dx$

and integration yields:

$\int d(-1/xy) = \int dx + c$

For which we obtain:

$-1/xy = x + c$

Or: $y = -1/x (x + c)$

Another more "refined" way to work with integrating factors starts with writing the typical first order linear DE as:

$$dy/dx + Py = Q$$

The name objective is to account for P and Q and also find the integrating factor, **r**.

One method for solving the DE shown is to find some function, usually $r = r(x)$ such that if the equation is multiplied by r, the left side becomes the *derivative of the product* ry. That is:

$$r(dy/dx) + rPy = rQ$$

and we then make the effort to impose upon r the condition that:

$$r(dy/dx) + rPy = d/dx (ry)$$

Which is not always easy, but can be if one is clever enough!

Expanding the right side of the previous eqn. via differentials:

$$d/dx (ry) = (rdy + y\, dr) / dx$$

and adding to the left, gives:

$$r(dy/dx) + rPy + (-r (dy/dx) - y (dr/dx))$$

$\Rightarrow dr/dx = rP$

If $P = P(x)$ is a known function, we can solve for r:

Viz. $dr/r = Pdx$ and $\ln r = \int Pdx + \ln C$

So that: $r = \pm\, C\, \exp(\int P\, dx)$

And C can be taken as: $C = \pm 1$

Then the function: $r = \exp(\int Pdx)$

Is called *the integrating factor*

Example: $dy/dx + y = \exp(x)$

$P = 1,\ Q = \exp(x)$

Then $r = \exp(\int dx) = \exp(x)$

So: $\exp(x)y = \int \exp(2x) + C = \exp(2x)/\,2 + c$

And $y = \exp(x)/2 + C\, \exp(-x)$ or:

$y = e^x/2 + Ce^{-x}$

Applications of Differential Equations

We now examine a number of distinct examples for applied differential equations.

Rate of flow: A 100 gallon tank is full of pure water. Let pure water run into the tank at the rate of 2 gals/ min. and a brine solution containing 1/2 lb. of salt run in at the rate of 2 gals/min. The mixture flow out of the tank through an outlet tube at the rate of 4 gals/min. Assuming perfect mixing, what is the amount of salt in the tank after t minutes?

Solution: Let s be the amount of salt in the tank in pounds at time t.

Then: $s/100$ = concentration of salt (i.e. as a proportion of total gallons of pure water in tank initially)

Therefore: ds/dt = *net rate* of change = (rate of gain in lbs/min - rate of loss in lbs/min)

We can further write:

$ds/dt = 1 - 4s/100 = 1 - s/25$

Writing the basic differential equation to solve:

$ds/(25 - s) = dt/25$

This requires integrating both sides:

$$\int_0^s -ds/(25 - s) = - \int_0^t dt/25$$

Note the integration is taken from 0 to s on the left side and from 0 to t on the right. This leads to:

$\ln(25 - s) - \ln(25) = -t/25$

And finally:

$s = 25 (1 - e^{-t/25})$

Let's take a time t = 25 minutes, what do we get?

$s = 25 (1 - e^{-25/25}) = 25 (1 - e^{-1}) = 25 (1 - 0.3678) =$

$25(0.6322) = 15.8$ lbs.

Spring system:

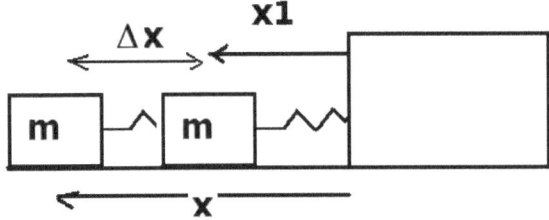

A block of mass m = 2.0 kg rests on a smooth horizontal surface attached to a spring. The spring has the property that it is stretched Δx = 0.05 m by a force of 10 N. If the block is displaced 0.05 m from the equilibrium position and released, find: the frequency and period of the motion.

Solution: We apply Newton's 2nd law of motion:

$F = ma = -kx$ where k is *the spring constant.*

Since F = 10N when x = 0.05 m, then:

$k = F/x = (10 \text{ N})/ (0.05\text{m}) = 200 \text{ Nm}^{-1}$

Now rewrite the force balance equation ma = -kx:

$m(d^2x/dt^2) = -kx$ or: $m(d^2x/dt^2) + kx = 0$

Divide through by the mass, m:

$(d^2x/dt^2) + (k/m)x = 0$

Which is the basic equation for the simple harmonic oscillator: $(d^2x/dt^2) + \omega^2 x = 0$

Where: $\omega = \sqrt{(k/m)} = \sqrt{(200 \text{ Nm}^{-1}/2 \text{ kg})} =$

$\sqrt{(100 \text{ Nm}^{-1}/\text{kg})} = 10 \text{ s}^{-1}$ or 10 radians/ sec

The frequency, f is related to angular frequency $\omega : \omega = 2\pi f$ so:

$f = \omega/2\pi = (10 \text{ s}^{-1})/2\pi = 5/\pi \text{ s}^{-1}$

The period $T = 2\pi/\omega = 2\pi/(10 \text{ s}^{-1}) = \pi/5 \text{ s} = 0.63 \text{ s}$

Projectile motion:

The layout of the projectile motion (missile)

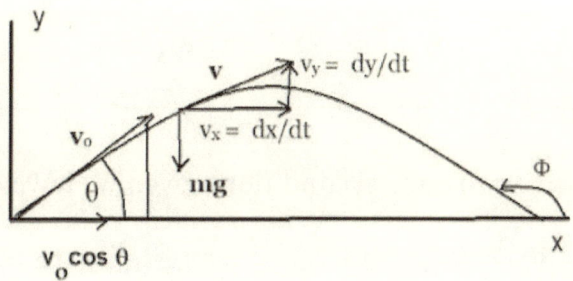

We write the DEs at t=0, for position: x=0, y = 0:

$$v_x = dx/dt = v_0 \cos(\theta)$$

$$v_y = dy/dt = v_0 \sin(\theta)$$

The force components at time t are:

$$F_x = 0 \quad \text{and:} \quad F_y = -mg$$

With these incorporated the differential equations become:

i) $m \, d^2x/dt^2 = 0$ and ii) $m \, d^2y/dt^2 = -mg$

Each equation above requires two integrations, yielding 4 constant of integration in all.

For eqn. (i) we get on integrating: $dx/dt = c_1$ Then:

$$\int dx = c_1 \int dt$$

Yields: $x = c_1 t + c_2$

Here for eqn. (ii) we have:

$d^2y/dt^2 = -g$ So: $dy/dt = -gt + c_3$

And: $y = -\frac{1}{2} g t^2 + c_3 t + c_4$

Based on the initial conditions given we have:

$c_1 = v_0 \cos(\theta)$, $c_2 = 0$, $c_3 = v_0 \sin(\theta)$, $c_4 = 0$

Then the position of the projectile at time t seconds after firing can be found from:

$x = v_o \cos(\theta) t$ and: $y = -\frac{1}{2} g t^2 + v_o \sin(\theta) t$

Clearly, the projectile attains maximum altitude when its y -component of velocity is 0, i.e.

$dy/ dt = 0 = -gt + v_o \sin(\theta)$

This must occur at time:

$t' = v_o \sin(\theta)/ g$

Then the maximum altitude is:

$y_{max} = -\frac{1}{2} g t'^2 + v_o \sin(\theta) t' =$

$-\frac{1}{2} g [v_o \sin(\theta)/ g]^2 + v_o \sin(\theta) [v_o \sin(\theta)/ g]$

$$y_{max} = v_o \sin(\theta)^2 / 2 g$$

Capacitor Discharge:

Whenever a capacitor C is discharged through a resistor R we observe the following behavior as the charge slowly decreases over time t:

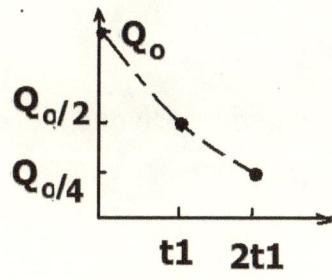

311

The potential difference across C is V. Therefore have: $Q = CV$. The current discharged can then be expressed:

$I = - dQ/dt$

(The charge, Q, *decreases* as the current I increases)

We know that $Q = CV$ but by Ohm's law: $V = IR$

So: $Q = -CR (dQ/dt)$

On separating variables we can integrate:

$$\int_0^t dt/CR = -\int_{Qo}^Q dQ/Q$$

Where:

$$\int_{Qo}^Q dQ/Q = \int_Q^{Qo} dQ/Q = \ln(Q_o/Q)$$

So that: $t/CR = \ln(Q_o/Q)$ or:

$e^{-(t/CR)} = Q/Q_o$

Finally: $Q = Q_o\, e^{-(t/CR)}$

Radioactive Decay:

A graph of generic radionuclide decay is shown below.

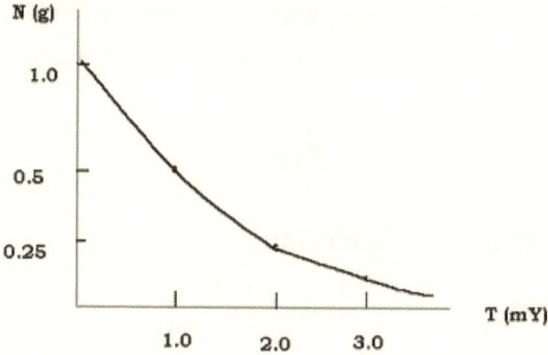

Here the vertical axis shows a relative scale for the amount or mass of some, unnamed decaying isotope which commences decay at some initial specified value, e.g. 1 gram then decreases to half that original amount in one half life:

This bears an analogous form to capacitor discharge.

We define the activity A of a radioactive source as:

$A = dN/dt = - \lambda N$

Where λ is the decay constant. The negative sign appended to the equation indicates that the amount N is decreasing with time t. Separate variables:

$dN/N = - \lambda dt$

We can write the integral:

$$\int_{N}^{No} dN/N = \ln (N_0/N) = - \lambda \int_{0}^{t} dt$$

313

The radioactive decay, then, is based on some original number of atoms N_0 decaying with an activity λ over time t, so:

$N = N_0 \exp(-\lambda t)$

The Infinite Square Well:

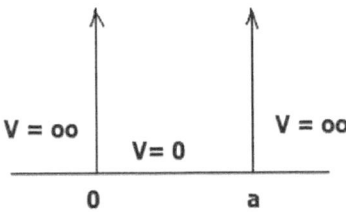

Probably the simplest quantum system of all is the 1-dimensional "infinite" square well shown above with accompanying limits from 0 to a. This is the x - direction (though granted it looks like y!) Inside the box the potential V = 0 and outside, V = ∞, hence the term, "infinite square well".

From these constraints, one sees that any particle cannot be at locations, x > a or less than 0, else it would have infinite energy. As we know, energies in QM are quantized, so no infinities are possible.

Therefore, outside the interval (0,a) one has the quantum wave form:

$\psi^*\psi = 0$

The Schrodinger wave equation is written (for one dimension):

$H^\wedge \psi = E \psi$

where $H^\wedge$ denotes the Hamiltonian operator:

$H^\wedge = [-i\hbar\, d/dx]\,^2/2m$

Then the Schrodinger equation to solve becomes:

$-\hbar^2/\,2m\,(\,d\psi^2/dx^2\,) = E\psi$

Thence:

$d\psi^2/dx^2 + 2m/\,\hbar^2\,(E\psi) = 0$

Or, in more concise form:

$d\psi^2/dx^2 + K^2\,\psi = 0$

where $K = \sqrt{[2mE]}\,/\,\hbar$

And this second order differential equation has the solution:

$\psi = C \sin Kx + D \cos Kx$

(C, D constants)

We set boundary conditions thus:

At $x = 0+$, $\psi\,(0+) = D$ (since $\sin(0) = 0$)

At $x = 0$, $\psi\,(0) = 0$

As a cautionary note, one must postulate that the

wave function is continuous (since we want ψ*ψ to represent something physically observable):

Then:

ψ (0+) = ψ (0-) therefore D = 0

and

ψ = C sin Kx

Now, at x = a+, ψ (a+) = 0

At x = a-, ψ (a-) = C sin K(a)

For continuity: ψ (a+) = ψ (a-)

Therefore: C sin K(a-) = 0   but C can't = 0, or no particle!

Therefore: sin K(a-) = 0

$\Rightarrow$ sin ($\sqrt{[2mE]}$ / ℏ) a = 0, and

(($\sqrt{[2mE]}$ / ℏ) a = n π

Therefore:

a = nπ/K and n = ± 1, ± 2, ..etc.

Thus:

2mE = $n^2 \pi^2 ℏ^2$

$$E = \{ n^2 \pi^2 \hbar^2 \} / 2ma^2$$

which yields the energy eigenvalues.

$$E = \{ n^2 \pi^2 \hbar^2 \} / 2ma^2 \; (E_n = 1, 2, 3, \ldots n)$$

Thus, E_n is restricted to:

$\pi^2 \hbar^2 / 2ma^2,$

$4 \pi^2 \hbar^2 / 2ma^2$

$9 \pi^2 \hbar^2 / 2ma^2 \quad$ etc.

We may also be concerned with the probability of where a particle is within a given system. The principle of probable position computation is based upon the wave function ψ itself not being a quantity one can measure. One instead measures: $\|\psi\|^2$

In general, the probability for a 3D system:

$$P = |\psi^2| \; dV$$

For a one-dimensional system: $P = |\psi^2| \, dx$

The probability P_{ab} of finding a particle between a and b:

$$P_{ab} = \int_a^b \|\psi\|^2 \, dx$$

APPENDIX:

(Frequently Used Physical Constants)

Boltzmann constant: $k = 1.3807 \times 10^{-23}$ JK^{-1}

Elementary electronic charge: $e = 1.6022 \times 10^{-19}$ C

Electron mass: $m_e = 9.1094 \times 10^{-31}$ kg

Proton mass: $m_p = 1.6726 \times 10^{-27}$ kg

Gravitational constant: $G = 6.6726 \times 10^{-11}$ m^{3}s^{-2}kg^{-1}

Planck constant: $h = 6.6261 \times 10^{-34}$ J s

$\hbar = h/2\pi = 1.0546 \times 10^{-34}$ J s

Speed of light in vacuum c 2.9979×10^8 ms-1

Permittivity of free space: $\varepsilon_0 = 8.8542 \times 10^{-12}$ Fm^{-1}

Permeability of free space: $\mu_0 = 4\pi \times 10^{-7}$ Hm^{-1}

Proton/electron mass ratio: $m_p/m_e = 1.8362 \times 10^3$

Electron charge/mass ratio:

$e/m_e = 1.7588 \times 10^{11}$ Ckg^{-1}

Rydberg constant $R_\infty = 1.0974 \times 10^7$ m^{-1}

Bohr radius: $a_0 = 2\ 5.2918 \times 10^{-11}$ m

Fine-structure constant $\alpha = 7.2974 \times 10^{-3}$

Stefan-Boltzmann constant: $\sigma = 5.6705 \times 10^{-8}$ $Wm^{-2}K^{-4}$

Wavelength associated with 1 eV: $\lambda_0 = hc/e = 1.2398 \times 10^{-6}$ m

Frequency associated with 1 eV: $\nu_0 = 2.4180 \times 10^{14}$ Hz

Energy associated with 1 eV: $h\nu_0 = 1.6022 \times 10^{-19}$ J

Avogadro number $N_A = 6.0221 \times 10^{23}$ mol^{-1}

Gas constant: $R = 8.3145$ JK^{-1}mol^{-1}

NOTES